Springer Series in Synergetics

Editor: Hermann Haken

Synergetics, an interdisciplinary field of research, is concerned with the cooperation of individual parts of a system that produces macroscopic spatial, temporal or functional structures. It deals with deterministic as well as stochastic processes.

Peter J. Plath (Ed.)

Optimal Structures in Heterogeneous Reaction Systems

With 115 Figures

Springer-Verlag Berlin Heidelberg New York
London Paris Tokyo Hong Kong

Peter J. Plath
Institut für Angewandte und Physikalische Chemie,
Universität Bremen, Bibliothekstrasse,
D-2800 Bremen 33, Fed. Rep. of Germany

Series Editor:

Professor Dr. Dr. h. c. Hermann Haken
Institut für Theoretische Physik der Universität Stuttgart, Pfaffenwaldring 57/IV,
D-7000 Stuttgart 80, Fed. Rep. of Germany and
Center for Complex Systems, Florida Atlantic University,
Boca Raton, FL 33431, USA

ISBN-13:978-3-642-83901-6   e-ISBN-13:978-3-642-83899-6
DOI: 10.1007/978-3-642-83899-6

Library of Congress Cataloging-in-Publication Data. Optimal structures in heterogeneous reaction systems / Peter J. Plath, ed. p. cm. – (Springer series in synergetics ; v. 44) Includes bibliographical references. 1. Chemical reaction, Rate of – Congresses. 2. Self-organizing systems – Congresses. 3. Heterogeneous catalysis – Congresses I. Plath, Peter, 1942–. II. Series.   QD502.068 1989 541.3'9 – dc20   89-21713

© Springer-Verlag Berlin Heidelberg 1989
Softcover reprint of the hardcover 1st edition 1989

2154/3150-543210 – Printed on acid-free paper

Preface

The concept of this book was developed during the Winter Seminar held in the Austrian mountains at the Alpengasthof Zeinisjoch, Tirol-Vorarlberg, from February 27 to March 3, 1988. Leading experts and advanced students in mathematics, physics, chemistry and computer science met to present and discuss their most recent results in an informal seminar. These were the circumstances that led to the idea of compiling some of the essential contributions presented at this seminar together with others describing basic features of "optimal structures in heterogeneous reaction systems".

The aim of this book is to present the scientific results of the intensive work carried out in each of the specific fields of research. Each contribution therefore presents the current state of the art together with a deeper treatment enabling a more comprehensive understanding of that particular field of work. The common ideas which unite all the different contributions are already expressed in the title of this book. The nature of heterogeneous reaction systems is quite varied. An example is provided by the chemical systems such as noble metal particles which may act as heterogeneous catalysts for gaseous chemical compounds. Under these circumstances the metal particles and/or their surfaces may undergo phase transitions during reaction. Imbihl and Plath report on special catalytic systems of this kind, which are of industrial importance.

But, as shown by Schuster, large and complex molecules in biochemistry and biological systems may also act as heterogeneous reaction systems, forming individual species that undergo chemical development in biological evolution.

In general, heterogeneity only requires a spatial division of the properties of matter and this can be achieved either by individual species or, more generally, by a non-continuous distribution of properties in space. Thus even liquid or gelatinous systems can be regarded as heterogeneous reaction systems if the reaction forces the system to organize itself by forming spatial structures which separate different reaction regimes from one another.

The question that produced the most discussion during the Winter Seminar was: how does a heterogeneous reaction system develop its optimal reaction structure by self-organization? Different ideas were developed. There is the concept of composing a system from a set of its subsystems. This concept is reported in detail in the article by Bestehorn et al. Highlighting the idea of networks, a similar approach is demonstrated by Ebeling et al. Accentuating the idea of heterogeneity, the work of Plath describes self-organization of the catalytic reaction system by means of cellular automata. Avnir et al. consider optimal reaction structures in geometrical terms, stressing the idea of a fractal organization. Zumofen et al. study the diffusion-limited reaction kinetics of simple chemical processes in distorted systems, introducing the hierarchy of an ultrametric space. They emphasize the idea that microscopic reactions create non-homogeneities by themselves. Schuster investigates the optimization problem with respect to the evolution of simple biological organisms using the methods of simulated annealing and genetic algorithms. In addition to all this work there is an article by Rössler on the explicit observer or the way we create different realities by describing them from the inside or outside. He favours the idea of a fully reversible description of the world that also creates all the wonderful dissipative patterns which can be observed at all levels of the universe. Addressing the fundamental question of how to introduce the concept of heterogeneity into time, he also arrives at a new idea of "Nowness".

I would like to thank all the authors of this book for their exciting contributions, and also all the participants of the 8th Winter Seminar at Zeinisjoch who stimulated the extensive discussions of such fundamental problems.

The organization of this meeting within the framework of interdisciplinary fundamental research would have not been possible without the generous support of the Office of Naval Research (Navy Grant N 00014-88-J-9013) and the Volkswagenwerk Foundation (Grant I/62 665). However, this book is neither a report nor the proceedings of the 8th Winter Seminar at Zeinisjoch, although it was inspired by this meeting. The publication of this book was mainly encouraged by Prof. H. Haken. The assistance of Dr. H. Lotsch and Dr. A.M. Lahee from Springer-Verlag is also gratefully acknowledged.

Bremen, July 1989   *P. J. Plath*

Contents

Modelling of Heterogeneously Catalyzed Reactions by Cellular Automata of Dimension Between One and Two

P.J. Plath

Institut für Angewandte und Physikalische Chemie, Universität Bremen, Bibliothekstraße NW 2, D-2800 Bremen, Fed. Rep. of Germany

Abstract. Experiments on the catalytic oxidation of CO on palladium supporting catalysts are modelled by the use of cellular automata. The discrete character of the models does not represent the molecular events, but reflects the microscopic origin of the macroscopic patterns that are observed. With one-dimensional automata it is mainly the temporal patterns of the oscillating reaction that can be modelled, whereas automata with a fractal dimension between one and two are used to study spatial pattern formation on an idealized model for amorphous catalytic material.

1. Introduction

Catalysis is one of the most fascinating concepts in chemistry. One cannot explain catalysis in terms of simple chemical mechanisms. Using the level of formal kinetic description, catalysis is characterized by a set of extremely non-linear ordinary differential equations. This implies that it could be necessary to describe a catalytic system by a variety of regimes which differ qualitatively in their dynamics. Because of the complexity of catalytic systems many quite different mechanisms have been suggested to explain experimental observations of a single catalytic system, and chemists can find very good reasons supporting each of these proposed mechanisms.

For industrial reasons heterogeneous catalysis has become a very important branch of catalysis. However, as far as technical catalysis is concerned, most investigations have been directed towards equilibrium state or fixed-point stationary behaviour close to equilibrium. Obviously, simple models are sufficient to fulfil the required constraints.

The well-known Langmuir-Hinshelwood mechanism and the Eley-Rideal mechanism are examples of the results that are achieved when dealing theoretically with heterogeneous catalysis in this way [1-5].

Investigating electrochemical systems which are related to heterogeneous catalytic systems. Bonhoeffer and Franck [6-10] first studied the complex dynamic behaviour of surface reaction systems. Although surfaces of electrodes are usually regarded as compact metallic surface planes, electrochemical processes such as etching and corrosion are often accompanied by the formation of local electrical elements, pitting and islands of reactive adsorbed species. Similar things happen in heterogeneous catalysis even though one uses fully connected metallic surfaces.

In the chemical industry, however, catalysts containing very small metal particles supported on inert, amorphous materials are much more important than compact plane and clean metal catalysts. Wicke and

1

Hugo [11-13] have detected the oscillatory properties of the H_2 and the CO oxidation reaction taking place on such industrial catalysts. There is a host of literature dealing with reactions catalysed by supported metals in which a variety of mechanistic effects are described such as the spill-over effect, contact catalytic effects and the influence of the supporting materials on the catalysis. Most of the corresponding models are based either on the stationary fixed-point behaviour or on the the ideal metallic surface, so far neglecting either the complex surface dynamics or the discrete character of the supported metal particles respectively.

There are several ways out of this unsatisfying theoretical situation. Ertl et al.[14-17] and Imbihl [18] investigated chemical adsorption and reactions on very clean and nearly ideal single crystal surfaces in the vacuum. They observed oscillations in space and time which could be correlated with the oscillating phase transitions of the two-dimensional adsorption phases of the covered surfaces. Schmitz et al. [19-22] looked for the heat production on macrsocopic metal surfaces and pellets of the amorphous support catalysts during the catalytic oxidation of H_2 and CO. They found local and temporal oscillating patterns. Both systems can be described using classical methods of continuous dynamics.

On the other hand, Jaeger and Plath [23-26] investigated metal catalysts especially palladium particles supported by amorphous material or incorporated in zeolite crystals. They concentrated their attention on the role of the metal particle size and the dilution of the metal particles inside the catalyst, in order to study particle size effects and coupling mechanisms in heterogeneous catalysis [27-29]. They proposed two types of model [30-32] to explain their experimental observations: first they introduced the storage model in which they enhanced the dimension of the catalytic reaction by an additional storage which undergoes phase transitions during reaction. Correlated to this idea, they proposed that the bulk of the metal catalyst may act as a storage for intermediate products, but there might indeed exist quite other storages. For theoretical reasons the chemical nature of the actual storage is of less importance for the analysis of the dynamics of such systems.

Hudson [33,34] also studied storage-aided systems and demonstrated the complex phenomenology of such systems.

Based on the use of cellular automata we have developed a second model taking into consideration the distribution of the discrete metal particles within the catalyst [24,32,35,36]. The elements of this model are mathematical cells of an automaton. These cells form the reactive units of the process and the state of the cells represents the actual degree of filling of the storages. In the theoretical analysis of this model great attention has to be directed to the coupling of the cells in order to obtain good correlations with the experimental reality.

In the following I shall report on different aspects of the use of such cellular automata for modelling heterogeneously catalised reactions.

2. Experimental Facts – First Essentials of Modelling

The experimental set-up has been described in detail elsewhere [23, 24]. So let me restrict myself to the few experimental facts which are of most importance for the understanding and the modelling of the system. In the experiments reported elsewhere [23,37-39], palladium single crystals are used to catalyse the oxidation of alcohols or carbon monoxide. The crystals are either incorporated in a crystalline framework of zeolites of type X (faujasite) or supported by amorphous Al_2O_3.

In the case of zeolite support the catalyst is a system of crystals within crystals. So the reactants have to pass through the zeolite channels which are of molecular size (0.8 - 1.2 nm), in order to get into contact with the surface of the palladium single crystals. But a very serious question is raised here: what do we mean when we speak about a single crystal surface which is surrounded by a zeolite crystal framework?

Even in case of the amorphous Al_2O_3 support, the surface of the palladium crystals is not really well defined. However, we can speak about the accessible surface. But its accessibility depends upon the molecular size and shape of the reactants as well as the rugged character of the supporting material. In any case, the reaction is accompanied by very complex transport and diffusion processes inside the catalyst material.

Moreover, about 20 mg of the zeolite crystals or the grains of the amorphous catalyst are distributed randomly on a disk or a sieve (ca. 1 - 2 cm in diameter) forming a nearly two-dimensional reaction layer. A gaseous flow of the reactants passes through the "shallow-bed" arrangement of the catalyst.

Using thermocouples, additional infrared measurements and mass spectroscopy the heat-production of the exothermic reaction and the temporal change in the product distribution can be continuously registered.

There are constraints in running the reaction in an oscillatory state or in a multistable regime. In the case of CO oxidation the sharp amplitudes of the oscillating conversion sometimes reach 90% of the total conversion of the reaction. This means that almost all the parts of the catalyst have to act in synchrony pulling down the conversion. But in any case the occurrence of oscillations requires coherent reaction of the catalyst.

To model this reaction we have to take into account that 20 mg of the zeolite catalyst consists of about 10^{15} palladium single crystals. Using amorphous supports the number of palladium crystals is in the same order of magnitude.

In the case of the zeolite crystals these metal particles are embedded in approximately 10^7 zeolite crystals assuming that the size of the zeolite crystals is of the order of 10 µm on average.

The zeolite crystals or the amorphous catalysts are often com- pressed to pellets and afterwards ground by a pestle and sieved. The resulting size of the catalyst grains is about 0.1 - 1 mm. These grains are distributed more or less evenly on the catalyst holder, forming a shallow bed.

So one has to decide on which level the reaction should be described and modelled. Taking for example the molecular level which is the usual one in ordinary chemistry, one would formulate a set of reaction equations. Chemists argue on this level when talking about catalytic active sites and geometric arrangements of the individual molecules bon-

3

ded to distinguishable surface sites [40-43]. But there is the question which I have mentioned before: how does one define a surface and distinguishable sites on it if the palladium crystals are surrounded by a zeolite framework for instance?

Nevertheless, chemists would answer that there are rules for formulating a set of kinetic equations starting with a given set of reaction equations. Without any doubt, one can establish such a system of ordinary differential equations, but there remains the question as to what the meaning of terms such as concentration is if we do not know the size of the surface? Furthermore, one should know the identity of the individual particles whose concentration is to be measured. Do we have molecules sitting on a metal surface like balls on a billard table, or do we have agglomerates of adsorbed species acting as single reactive units?

One can, however, describe the reaction in this way if one takes a much more general view. One can span a continuous space by a set of variables and follow possible trajectories of the system within this space. However the question remains as to how to form those variables ab initio from the individual molecules of the reactants in the gaseous state. No doubt the siaving principle will allow us to describe these systems using such global and dominant modes. [44-46]. But in doing so, we change our level of description - it is no more the molecular level.

However, one can model the catalytic system on a microscopic level using the palladium single crystals as the basic units of the description. The earlier interpretations of the storage model given by Jaeger and Plath [24,37,47,48] are based on this level of description.

The continuous counterpart of the 10^{17} palladium particles is the phase of the palladium crystals. In the case of the CO oxidation the above authors assumed a phase transition between Pd and PdO as the essential process which causes the oscillations, because this process is highly non-linear. The bulk of the palladium crystals act as a storage and there is a threshold in its correlated filling rate decribing both the Pd and the PdO phases.

It has been supposed that CO and O_2 molecules are only adsorbed on the surface of the Pd phase. This should be the active state of the catalytic metal particles. Since CO is oxidized to CO_2, this reaction is connected with strong heat production which is local with respect to the single metal particles. The heat production now favours the oxidation of the palladium single crystals resulting in the PdO phase of the crystals. But in the PdO state no more O_2 can be adsorbed, so that the reaction $CO + 1/2\ O_2 \rightarrow CO_2$ stops. Now the surface of the PdO phase can be totally covered with CO which may reduce PdO to Pd inducing a phase transition that brings the system back to the metallic Pd phase [49].

The problem with such a means of modelling is the uncertainty in the properties of the phases, due to of the smallness of the Pd and PdO particles. However, the existence of both states in the oscillating regime is proved by X-ray measurements [24,49]. A solid phase is characterized by the existence of a surface but if the solid particles are very small, the properties of the surface become very important, creating an additional new phase - the surface phase. As a consequence, physical and chemical properties change drastically when proceding from the compact metallic state to microscopic particles of 1 - 10 nm in diameter. So when talking about phase transitions one is really using a formal description on a higher level which can be micros- copically interpreted. But in fact the terminology of phases does not really refer to this microscopic level of description.

There is another level on which to describe the reaction that is based on the zeolite crystals or corresponding agglomerates of the amorphous catalysts. On this level modelling has to take into account internal and external diffusion of the reactants within and among the zeolites. In the case of oscillations all zeolite crystals have nevertheless to act in synchrony. And there are still about 10^7 zeolite crystals available. Although one can measure these different diffusion coefficients [50-53], it is beyond the scope of a complete mathematical description to model the observed reaction diffusion system in this way.

On a larger scale one can consider agglomerates of zeolites or grains of the amorphous materials. One can estimate their number as $10^2 - 10^3$ within the reactor. These elements of the catalyst are not well defined with respect to their geometry and distribution. They are somehow arbitrarily chosen portions of the catalyst which are more or less distributed by chance. They are constructed because of the demand for elements representing a fully synchronized part of the catalyst.

It is now possible to model the reaction taking into account all of these individual catalyst parts providing one can specify the relations among them. This level of description will be the one for the discrete modelling of the reaction with the help of a cellular automaton.

There still remains the question, relating to the terms of the description. If one uses words like molecules or phases, are they still defined in the common way? In order to answer this question let me briefly refer to the ancient ideas about atoms and molecules introduced by Dalton and Avogadro [54,55]. Both involve the discrete description of gases.

Taking over their ideas seriously, one may come to the conclusion that modern chemistry is a discrete description of reactions, but not of homogeneous substances. The smallest, minimal set of units M_i of the i-th reaction forms its correlated set of molecules [56]. This idea can be generalized. In case of gaseous reactants and products these generalized molecules are identical to the idea of a molecule in the common sense. Structures representing the possible reactions can be correlated to these molecules. The reaction itself is the change of the molecular structure. This change is temporally bounded by the final states of the reaction. The length of the time interval of the reaction is usually normalized and the name of the interval is the reaction coordinate or the reaction pathway.

Ugi [57,58] has introduced the idea of an ensemble of molecules belonging to a family of molecules in order to describe the structural changes of a set of generalized molecules within the temporal unity of the set of reactions. He also introduced a metric with the idea of a chemical distance between reactants and products.

If, in addition, the reaction or a set of reactions is also locally bounded, it forms a reactor. One may generalize this idea and arrive at molecules, reactions and reactors on all levels of description. This would represent a kind of renormalization. Or speaking without group theoretical background, this is a way of creating new elementary units representing some scaling behaviour of the system. The idea of a cell is the most abstract and simple way of modelling an elementary chemical reactor. However, the state of a cell is the "channel" in which the reactor runs; it is its way of "producing". The structure of the correlated molecules is now very generally conserved in the state of the cell. So there are no longer any problems in adopting the desired level, sufficient to describe the experimental observations.

To model the macroscopic temporal patterns of the oscillating catalytic CO oxidation one may take reactors as the units of the description, and this may allow us to derive the essential variables of the system under consideration.

In adopting this kind of description of processes, the following problems arises: how does one define the states of the cells; detect their neighbouring cells; and establish their intercellular relations? From a knowledge of these features, the discrete temporal development of the state $s_i(t)$ at time t of each cell i could be derived in terms of the states $s_{ij}(t)$ of the cells ij neighbouring cell i :

$$s_i(t + 1) = F(s_{ij}(t), s_i(t)) \quad ; i,j,t \in \mathbb{N} \tag{1}$$

This last question can only be answered for simulations of very concrete systems. In most cases one has to put up with severe but very general simplifications which nevertheless bring fundamental understanding of the reaction. In the following section I shall present some of these simulations.

3. Simulation of the Heterogeneous Catalytic CO Oxidation by a One-Dimensional Automaton

Figure 1 shows a very typical example of the temporal patterns which occur in the oscillating regime of the heterogeneously catalyzed oxidation

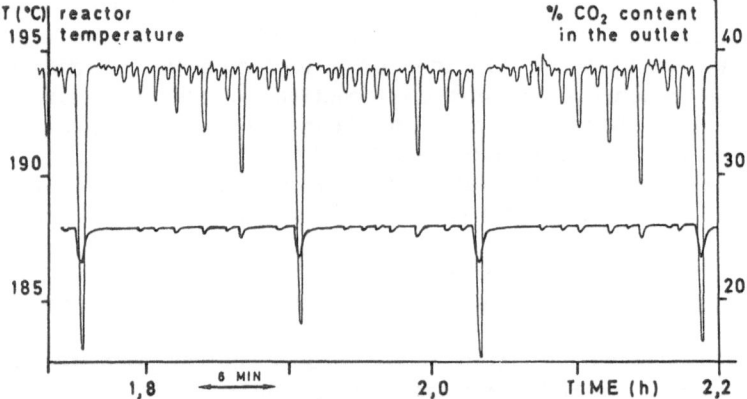

Fig. 1: Self-similar oscillations of the CO_2 production during the catalytic oxidation of CO using synthetic air and a Pd crystal loaded zeolite catalyst (20 mg catalyst, 0.38 vol % CO in the feed, temperature of the reactor T = 187,8°C. flow rate 9 l/h, volume of the reactor 14 ml, Pd content in the catalyst 14.7 wt %).

of CO. The picture creates the impression that to some extent the patterns seem to be self-similar and, on the other hand, periodic. Apparently this fact is of some importance for modelling this system.

The other observation, essential for the modelling is the sudden and total breakdown of the conversion. This means that almost all unit reactors of the catalyst will stop their production at the same time, or rather in the same small time interval.

For modelling, let me take into account only these few facts and let me take a very simple one-dimensional binary automaton which is directed to only one side. Thus, because of the binary character the state of a cell is either productive (white coloured cell) or blocked, i.e. a non-productive (black coloured cell). The relevant neighbourhood of a cell is only composed of itself and the lower (or left) neigh- bouring cells. The following simple evolution rules take into account only nearest neighbours:

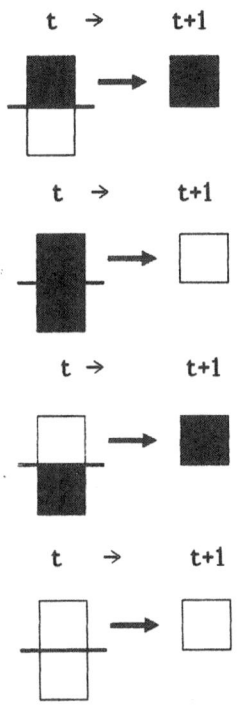

t → t+1   – a cell inactive at time t remains inactive at time (t+1) if its lower neighbour is active at time t;

t → t+1   – a cell, inactive at time t becomes active at time (t+1) if its lower neighbour is inactive at time t
– – this way the chemical reduction step is modelled to some extent;

t → t+1   – a cell, active at time t becomes inactive at time (t+1) if its lower neighbour is inactive at time t;
– – modelling of the oxidation step;

t → t+1   – a cell, active at time t remains active at time (t+1) if its lower neighbour is active at time t.

The sudden breakdown of the conversion means that a large number of reactors do not produce CO_2 any more within the same time interval. I do not as yet know anything about the spatial distri- bution of these reactors in the non-productive state. Let me sum the number of blocked cells at any time step. This sum Σ should be a function of time $\Sigma = \Sigma(t)$. The plot of the graph of this function approximates very well the experimental time series.

The visual similarities can be increased if one does not start with only one black cell but with several, arbitrarily distributed (see Fig. 4 and 5). This way the strong regularities are omitted although the self-similarity of the temporal patterns still remains.

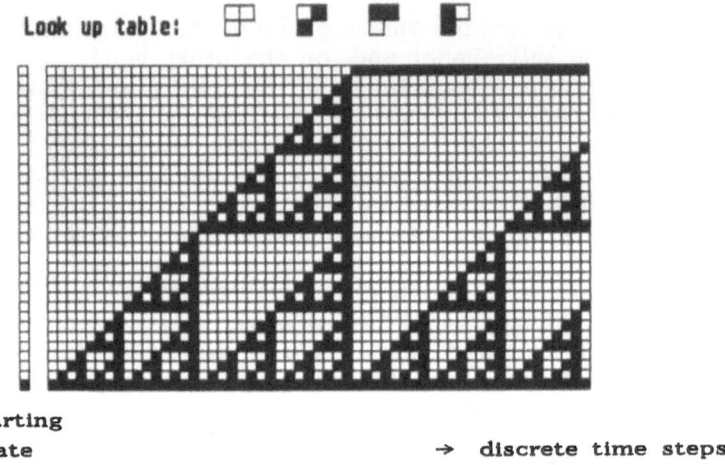

Fig. 2: Spatial and temporal development of the one dimensional cellular automaton described above 1,2-rule. This rule will produce a sequence of states of the automaton in which one isolated blocked cell at time t is followed by two blocked cells at time t+1.

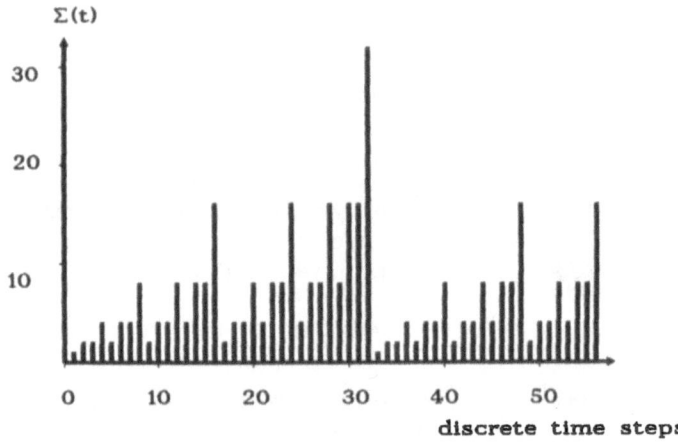

Fig. 3a. Time series of the sum $\Sigma(t)$ of the states s(t) for the 1,2-cellular automaton rule which is defined above

To prove whether there are other similarities between the simulation and the chemical experiment apart from the visual impression, let me map the sum $\Sigma(t+1)$ at time (t+1) as a function of the sum $\Sigma(t)$ at time t (Fig. 6). Comparing this graph with the Poincare map of the experimental time series, one is surprised at the congruity of the two maps. The automata that correspond with the experiments very well will be those whose evolution rules for the oxidation are based not only on nearest neighbours but also on more distant neighbours. This way the map approaches a rectangular shape as in the experimental Poincare maps (Fig. 7).

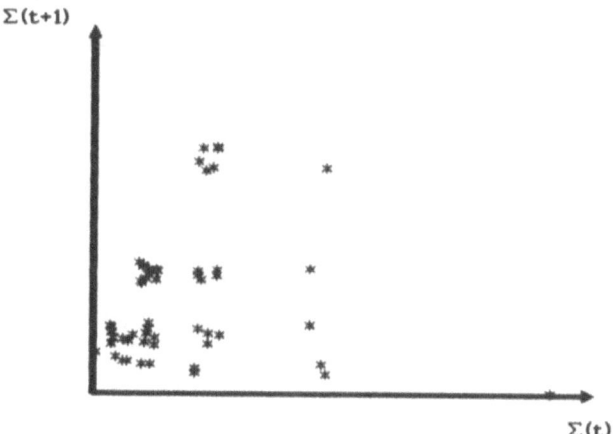

Fig. 3b. Poincare plot of the time series which is shown in Fig. 3a. In case of the discrete time series of the cellular automata this plot resembles the next-amplitude plots of the experimental time series

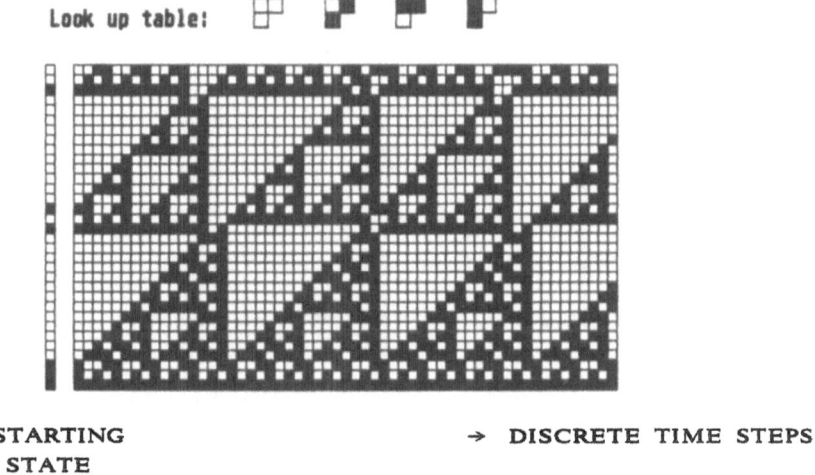

STARTING   → DISCRETE TIME STEPS
STATE

Fig. 4   Spatial and temporal development of a one-dimensional cellular automaton (1,2-rule) starting with a randomly chosen distribution of states s(0)= 1

But in the experimental Poincare maps one can also observe quadratic structures for small amplitudes of the oscillating breakdown of the conversion. This time series corresponds to lower concentration of CO in the inlet of the reactor. These patterns should be simulated by an automaton that considers not only the actual state of the cells but also their phase [59]. To some extent the evolution rules should also reflect the history of the cells especially if not only nearest neighbours are to be regarded (Fig 8). With such "historic" automata one is able to model

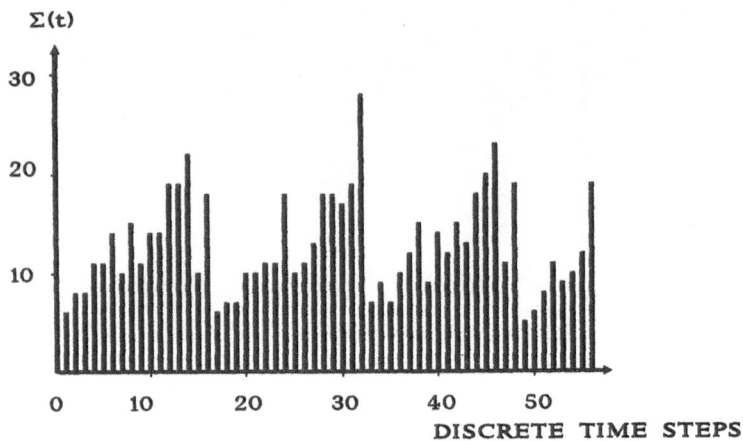

Fig. 5 Time series of the sum $\Sigma(t)$ of the states $s(t) = 1$ which is correlated to Fig. 4.

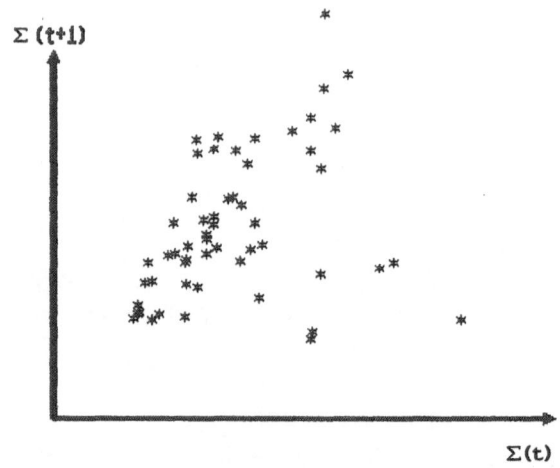

Fig. 6 Poincare map $\Sigma(t+1) = f(\Sigma(t))$ of the one-dimensional 1,2-rule cellular automaton, as shown in Fig. 4

the retarding effects while returning to the productive state of the unit reactors of the catalyst.

Another question concerns the periodicity within the experimental results. The periodic behaviour is usually investigated by Fourier analysis [25]. But in the model there is no periodicity because of the dilatation symmetry within the fractal Sierpinsky structures. However, if one introduces a finite length of the automaton, its temporal sequence becomes periodic. One can study the Fourier spectra of the automata by considering the function $\Sigma = \Sigma(t)$. The main frequency is correlated to the small but very fast oscillation. However, the pattern of the time series is globally structured by the subharmonics of this main frequency, but not by this frequency itself (Fig. 9).

Look up table:
STARTING
STATE
→ DISCRETE TIME STEPS

Fig. 7a. Spatial and temporal development of a one-dimensional cellular automaton with a 1, 3-rule

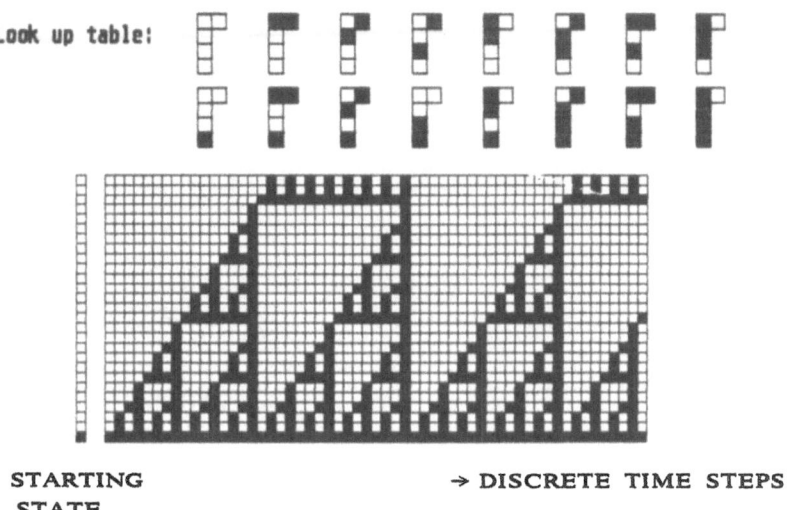

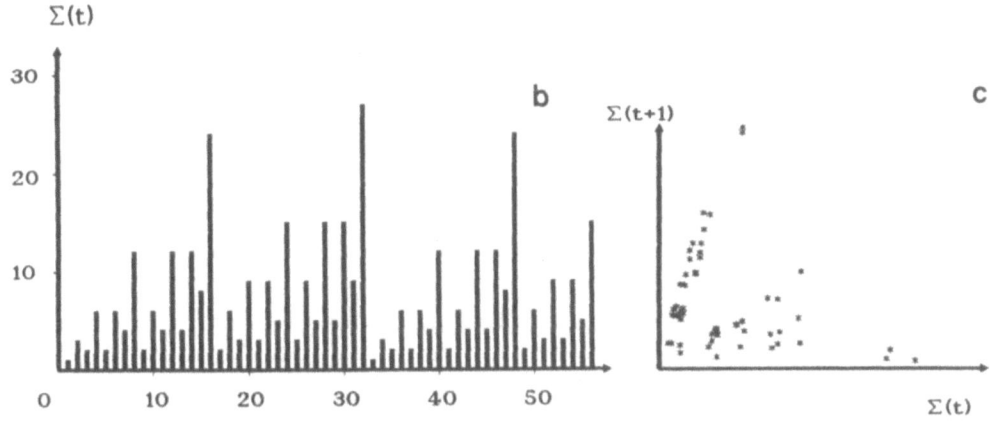

Σ(t)
30
20
10
0
10
20
30
40
50
b
c
Σ(t+1)
Σ(t)
DISCRETE TIME STEPS

Fig. 7b Time series of the 1,3-rule automaton

Fig. 7c Poincare map $\Sigma(t+1) = f(\Sigma(t))$ of the one-dimensional 1,3-rule automaton

This observation correlates with the experimental fact that the important visual structure of the time series of the chemical reaction is also based on the subharmonics, but not on the main frequency itself [25] (Fig. 10). It seems to me that this result is of much greater and general importance than this special case alone might suggest.

Although one can simulate a lot of different experimental patterns using one-dimensional automata, I do not claim, that one can model all aspects of the real chemical system in this way. But I do claim that

Look up table:

automaton
length l = 24

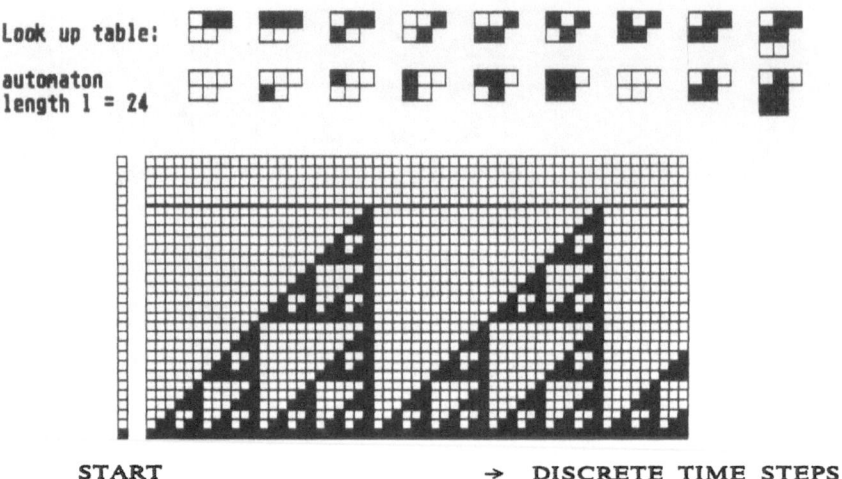

START   → DISCRETE TIME STEPS

Fig. 8 Spatial and temporal development of a one-dimensional cellular automaton governed by a 1,2,3-rule. The automaton contains 24 cells (length l of the automaton)

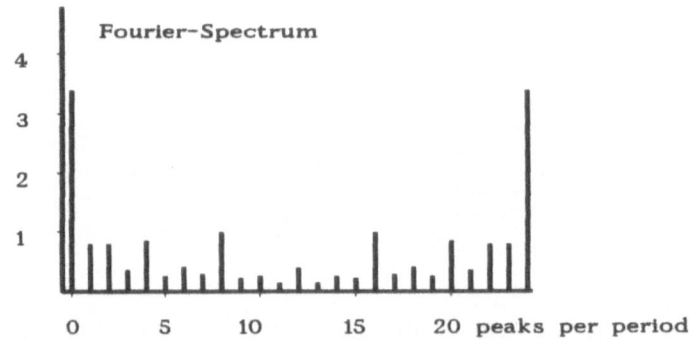

Fig. 9 Fourier spectrum of the one-dimensional 1,2,3-rule automaton shown in Fig. 8. Because of the period length of the time series 24 frequencies have been used

these automata are suitable for modelling the main features of the observed patterns of the experimental time series.

That is what I was looking for, namely a formalism that creates patterns similar to the experimental ones. Perhaps one may learn something about the chemical system through playing with such models since some features of the automata can be chemically interpreted. But that is up to the reader. On the other hand, there might be other models, even other automata, that accentuate other aspects of the experimental system.

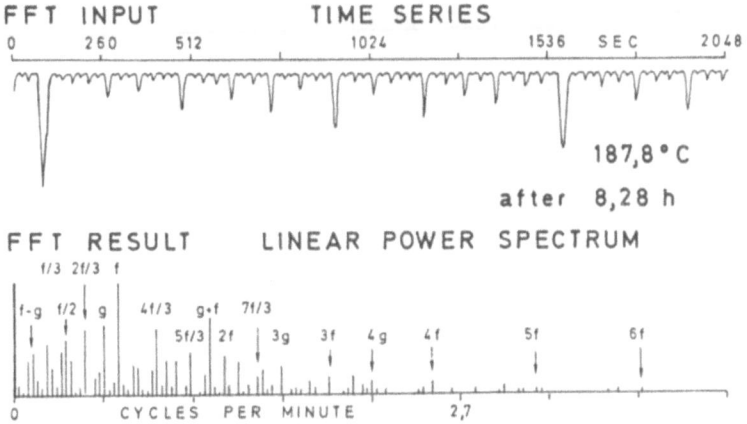

Fig. 10a Time series and linear Fourier spectrum of the oscillating catalytic oxidation of CO. (For further constraints of the reaction, see Fig. 1.)
This pattern arose 8.28 h after starting the reaction.

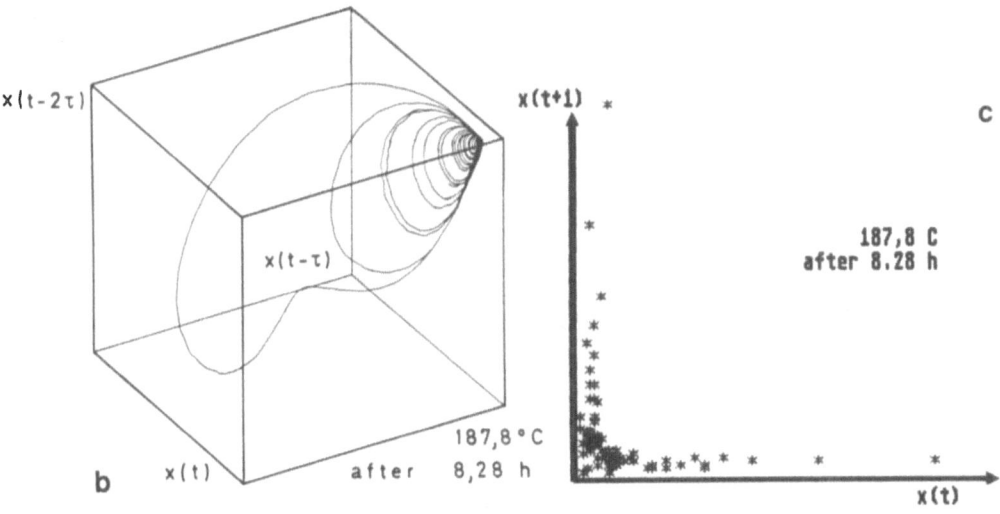

Fig. 10b Phase-space trajectory of this oscillating reaction; time delay τ =
3 s

Fig. 10c Poincare-plot of this experimental time series.

4. A Cellular Automaton Model of Fractal Dimension

I have not as yet discussed the fact that the zeolite crystals, or the
grains of the amorphous catalysts, are arbitrarily distributed on the
catalyst holder. There is no evidence that these particles are uniformly
distributed. Even the palladium particles within the mineral supports
might not be homogeneously distributed (Fig 11).

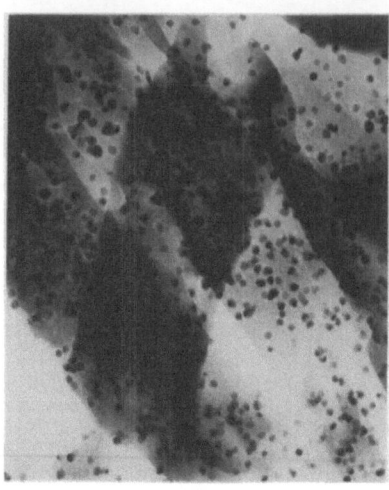

Fig. 11 Electron micrograph of the catalyst used for the catalytic experiments shown in Figs. 1 and 10.

Although I should like to restrict myself in the level of description to the elementary reactors formed by macroscopic parts of the catalyst, I am not convinced that their experimental arrangement is uniform. So, making allowance for the possibility of a nonuniform distribution of the elementary reactors, I investigated a fractal cellular automaton the framework of which is of a dimension close to D = 1.58 as shown below :

Table 1 Structure of the arrangement of cells in the fractal cellular automaton

number of cells C in the row										
2^0	C	0	0	0	0	0	0	0	0	0
2^1	C	C	0	0	0	0	0	0	0	0
	C	0	C	0	0	0	0	0	0	0
2^2	C	C	C	C	0	0	0	0	0	0
	C	0	0	0	C	0	0	0	0	0
	C	C	0	0	C	C	0	0	0	0
	C	0	C	0	C	0	C	0	0	0
2^3	C	C	C	C	C	C	C	C	0	0
	C	0	0	0	0	0	0	0	C	0
	C	C	0	0	0	0	0	0	C	C

By the term dimension I mean the fractal growth rate dimension of the cellular automaton [60]. This idea of a dimension refers to an embedding of a sequence of states into a discrete lattice. In this way, a set of cells, each cell denoted by capital C (Tab. 1) might be imbedded in a quadratic lattice forming a pattern $P(2^k)$ where $N(2^k)$ equals the sum of all cells up to the row number 2^k.

Doubling of the number of rows will give $N(2^{k+1}) = 3\, N(2^k)$ cells. For $k \to \infty$ one can compare the growth rate of the number $N(2^k)$ of cells with the growth rate of the number 2^k of rows.

$$N(2^k) = 3^k = 2^{k\, \text{ld}\, 3} \tag{2}$$

$$\frac{N(2^k)}{2^k} = \frac{2^{k\, \text{ld}\, 3}}{2^k} = \text{ld}\, 3 = D \approx 1.58 \tag{3}$$

This ratio d is called the growth rate dimension of the automaton. This arrangement of cells resembles a discrete version of the Sierpinsky gasket. Let me therefore call it a Sierpinsky lattice. For the limit of infinity each vertex of the Sierpinsky gasket has four nearest neighbours. This then resembles the von Neumann neighbourhood in an infinite square lattice.

If one octahedrally links a finite Sierpinsky gasket with itself the degree of each vertex becomes four too (Fig. 12). For a large number of cells the dimension of the finite and plane lattice goes to the fractal dimension $D = \log_2 3 \approx 1.58$ of the Sierpinsky gasket. Comparing the patterns of the Sierpinsky lattice and the Sierpinsky gasket, which is regarded as a simplicial complex of line segments, the cells can be mapped onto the parallel line segments of only one of the three possible directions of the gasket.

It would be worthwhile to find a lattice of the same dimension which could be mapped onto the vertices of the Sierpinsky gasket. One useful result of such a lattice is the existence of a well-defined nearest neighbourhood for each cell which is of degree four. Using a special transition rule called the "1,2,3,4,5-rule" one can create such a discrete pattern with a one-dimensional automaton. The growth rate dimension of this automaton is just $D = 1.584$. For the construction of

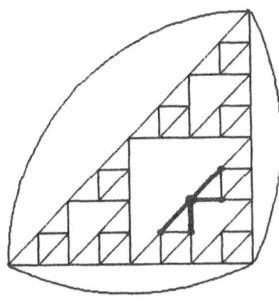

Fig. 12 An octahedrally closed graph of the Sierpinsky gasket

Sierpinsky lattice
1,2,3,4,5-rule

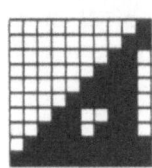 

Fig. 13 Spatial and temporal development of a one-dimensional automaton which is described by a historic 1,2,3,4,5-rule.

the transition rules one needs almost three nearest neighbours and also their history back two time steps (Fig. 13).

Let us also take this lattice as a model for the non-uniformly distributed elementary reactors within the catalyst. This lattice will represent all kind of pores and channels which might exist in a real catalyst. Moreover, it models the different connections among loosely as well as densely packed catalyst grains and elementary reactors. One can vary the range of interactions among the catalyst's parts by taking the connectivity distance as a measure for counting the num- ber of cells in the neighbourhood of any chosen cell with respect to the graph theoretical distance in the simplicial complex of the Sierpinsky gasket.

Taking into consideration a sufficiently large number of cells, the states $z_i(t)$, i = 1, 2, 3, ... which govern the states $z_i(t+1)$ of the i-th cell, one can observe strong oscillations going from t to (t+1) among the number of cells with the state $z(t) = 1$ as well as for the number of cells which change their state $z(t) = 1$ to $z(t+1) = 0$.

As one can see in Fig. 14, there is a drastic change of structures within the patterns which are correlated to the oscillations. In order to quantify this qualitative statement, one can calculate the growth rate dimension of all states of the automaton. Although there is only a minor oscillatory change in the avergage "coverage" of the automaton, one can observe a pronounced oscillation of the growth rate dimension, correlated to the stepwise transformation of the automaton (Fig. 15). The whole automaton is somehow synchronized like a "blinker" of its states changing between its two states all the time.

This interesting behaviour is due to the fact that during the state change ("Zustandswechsel") almost all cells of state $z(t) = 1$ are transfomed into the state $z(t+1) = 0$ whereas the reverse change $z(t) = 0 \rightarrow z(t+1) = 1$ forms at time (t-1) and also at time (t+1) something like a photographic negative of the pattern of states $z(t) = 1$ at time t.

As we know from two-dimensional automata there are some rules which will produce spatial patterns like circles or spirals moving across the arrangement of cells [61,62]. For the creation of such patterns one needs a dynamic system of at least three possible states of the cells which are passed via a cyclic flow [63-67]. That is true if the history of the cells is not involved. But the third state, the so-called refractory state can be replaced by the past state $z(t-1)$ of a cell in its present state $z(t)$.

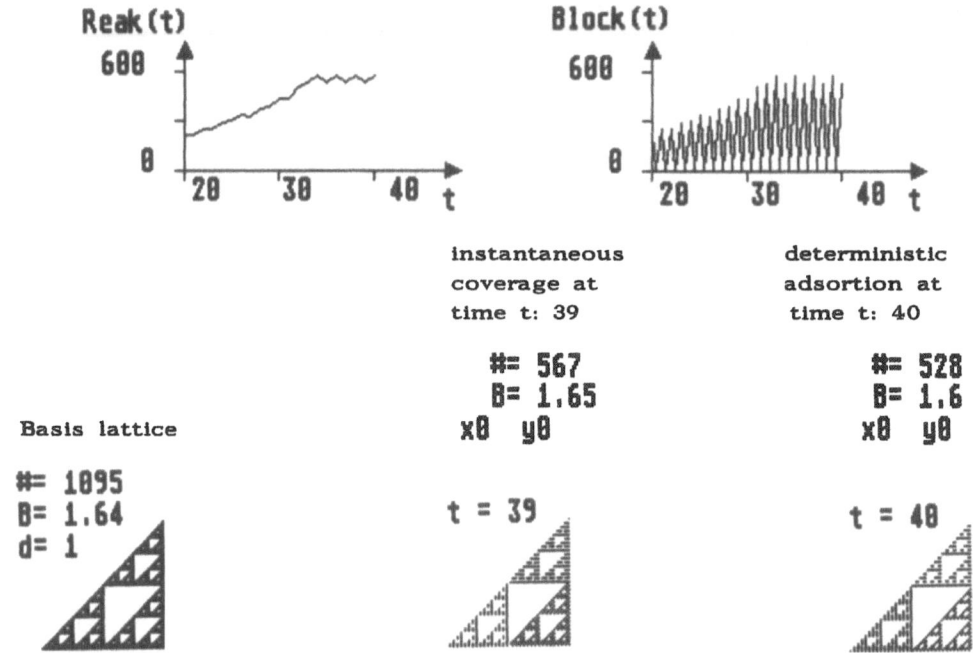

Fig 14 Oscillations of structures during the development of a fractal automaton after twenty time steps. The function Block(t) gives the number of cells with the state z(t) = 1 whereas the function Reak(t) represents the number of cells which are changing their state z(t) = 1 to z(t+1) = 0. At time t = 33 the automaton reaches a limit cycle. Only the four nearest neighbours are considered. One time step is divided into two parts: reaction and adsortion which correspond to the changes in state from 1 to 0 and from 0 to 1 respectively. The thresholds for these partial processes are #(z(t) =1) =1 and #(z(t) = 1) ≥ 1 for reaction and adsortion respectively.

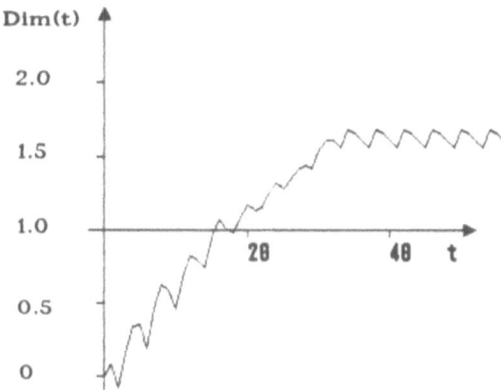

Fig. 15 Oscillation of the growth rate dimension during the development of the fractal automaton shown in Fig. 14. This dimension is calculated from the distribution of cells in the state z(t) = 1.

So let me assume that the past state will retard the state change ("Zustandswechsel") from $z(t) = 1$ to $z(t+1) = 0$. This means:

$$\text{for } z(t) = 1 \quad z(t+1) = \left\{ \begin{array}{ll} 1 & \text{if } z(t) = 1, \text{ and } z(t-1) = 0 \\ 0 & \text{if } z(t) = z(t-1) = 1 \end{array} \right. \tag{4}$$

$$\text{for } z(t) = 0 \quad z(t+1) = \left\{ \begin{array}{ll} 1 & \text{if } z(t) = 0 \text{ and } z(t-1) = 0 \\ 0 & \text{if } z(t) = 0 \text{ and } Z(t-1) = 1. \end{array} \right. \tag{5}$$

This rule reflects a hysteresis-like behaviour of the elementary reactors in the catalyst. If one now searches for patterns which resemble circles in the two-dimensional plane, one has to take into consideration the specific geometry of the Sierpinsky lattice. Within this framework a circular wave degenerates to two straight lines which are somehow orthogonal. If such a straight line reaches the exposed vertices of the lattice it creates in general two new lines which are orthogonal both to each other and to their mother lines.

A circle on this framework is defined by the set of all vertices which can be reached by a path of minimal length, which resembles the radius in the two-dimensional Euclidian space. This way a quadratic lattice will create a square-like circle.

Figure 16 demonstrates the spreading of a "circular" wave without dispersion like a soliton on a Sierpinsky lattice [67]. An annihilation can be observed if two orthogonal waves come into contact.

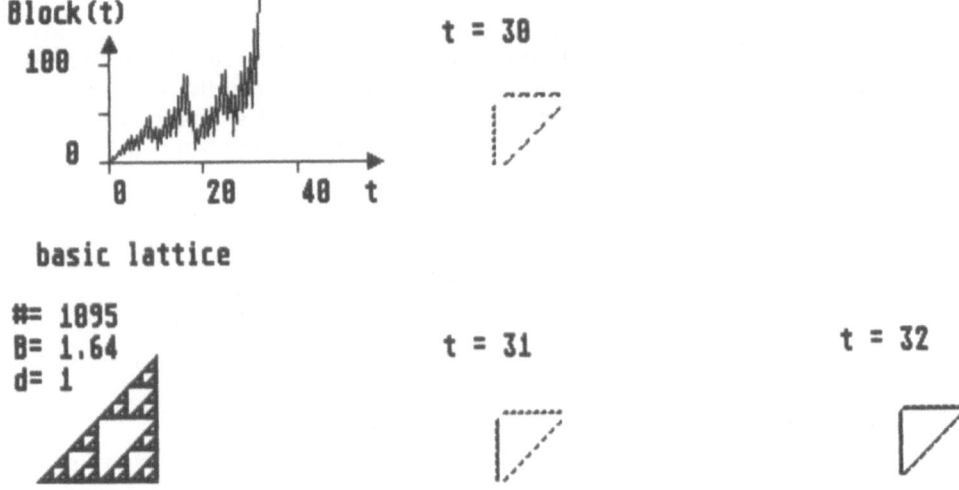

Fig. 16 Spreading of a "circular" wave on a Sierpinsky lattice. The starting point of this wave was the point (0,0) which is the left corner of the lattice. The "circle" consists of three "lines". These lines annihilate at the time step t=34. The historic rule of this automaton is based only on nearest neighbours and one time step is split into two parts. Reactive desorption takes place at time t if two or more neighbouring cells are in state 1 and the state of the cell of interest at time (t-1) is also 1. Adsorption happens if one or more neighbours are in state 1 and the state of the cell which is considered at time (t-1) equals 0.

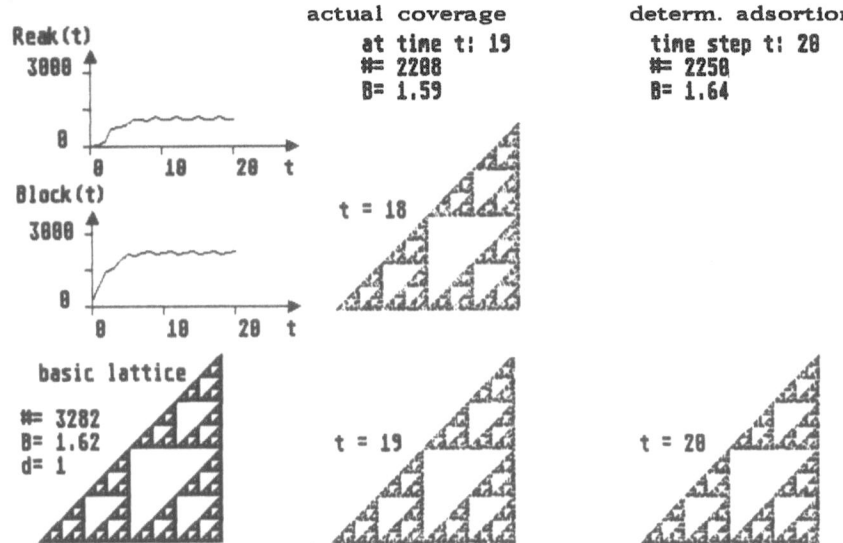

RULE: Reak ≥ 2, Ads ≥ 1

actual coverage
at time t: 19
#= 2288
B= 1.59

determ. adsortion
time step t: 28
#= 2258
B= 1.64

Reak(t)
3888
8
8   18   28   t

Block(t)
3888
8
8   18   28   t

t = 18

basic lattice
#= 3282
B= 1.62
d= 1

t = 19

t = 28

Fig. 17 Incessant birth of spreading "circles" on the basic Sierpinsky lattice The development of the automaton starts with a random distribution of cells with z(0) = 1. The historic rules of the automaton are described Fig. 16. The values reak and ads mean the numbers of neighbouring cells in the state z(t)=1 and z(t)=0, which are neaded for successful reactive desorption (z(t)=1 → z(t+1) =0) and deterministic adsorption (z(t) = 0 → z(t+1) =1) respectively. # is the number of cells in the basic automaton or the number of cells in the state z(t)=1 in the instantaneous state of the automaton respectively. B is the growth rate dimension and d is the distance between a cell and its nearest neighbours. The functions Block(t) and Reak(t) (see also Fig.14) reach oscillatory states.

Starting with a stochastic coverage of the Sierpinsky lattice oscillations of the coverage as well as of its correlated growth dimension can be observed (Figs. 17, 18). The solitary travelling circular wave is due to the "historic" rule, whereas one gets an incessant birth of "circles" without recourse to the state z(t-1) at time (t-1) if one starts with a stochastic coverage.

Comparing the results of these simulations with the chemical experiments on CO oxidation, one gets the impression that a single "circular" wave travelling across the Sierpinsky gasket models best describes the reality of the self-similar sequences of breakdowns of the conversion at relatively "low" temperature and low CO concentration in the feed [24]. But one has to recognize that one always needs at least one additional time step for the reduction. To model this fact, one has to distinguish between the two time steps "reactive desorption" and "adsorption". Within the framework of the chosen description this means that the reduction of the area of oxidized palladium which forms the unproductive set of elementary reactors needs some time. On the other hand, the spatial pattern of the blocked reactors seems to produce in its frontier surroun-

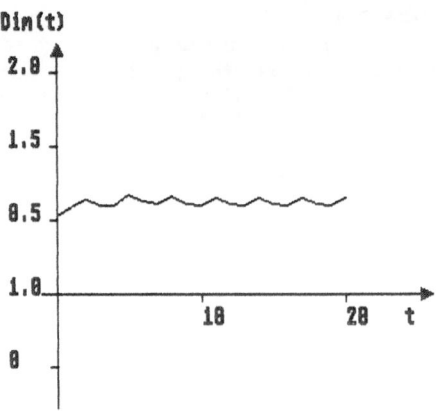

Fig. 18 Oscillation of the growth rate dimension correlated to the development of the automaton shown in Fig. 17.

dings a growing copy of itself. Temporarily separating into two time steps what happens in the previously described automata during only one time step, both the transitions $z(t) = 1 \rightarrow z(t+1) = 0$ and $z(t) = 0 \rightarrow z(t+1) = 1$ will create an interesting set of automaton rules which can be interpreted as follows (see Fig. 19):

$z_i(t) = 1 \rightarrow z_i(t+1) = 0$

if there exists at least one nearest neighbour (ij) the state of which is $z_{ij}(t) = 1$;

All clusters of non-productive reactors become active within the next time step.

$z_i(t) = 0 \rightarrow z_i(t+1) = 1$

if $z_i(t-1) = 0$ and $z_i(t-3) = 0$ and if there existed at least one nearest neighbour (ij) the state of which was $z_{ij}(t-1) = 1$.

Indeed, this special rule describes the formation of an invisible a priori state $z_i(t) = 0$ which evolves to $z_i(t+1) = 1$ at time (t+1) depending upon the size of the clusters (number of neighbouring cells) at times (t-1) and (t-3).

Let me compare this rule with the correlated set of rules for the one-dimensional 1,2,3,4,5-rule automaton. This rule takes into consideration the three low nearest neighbours and the history of the relevant cell and its neighbours back to the time (t-2). As one can see, this rule is highly complex in its spatial and temporal structure.

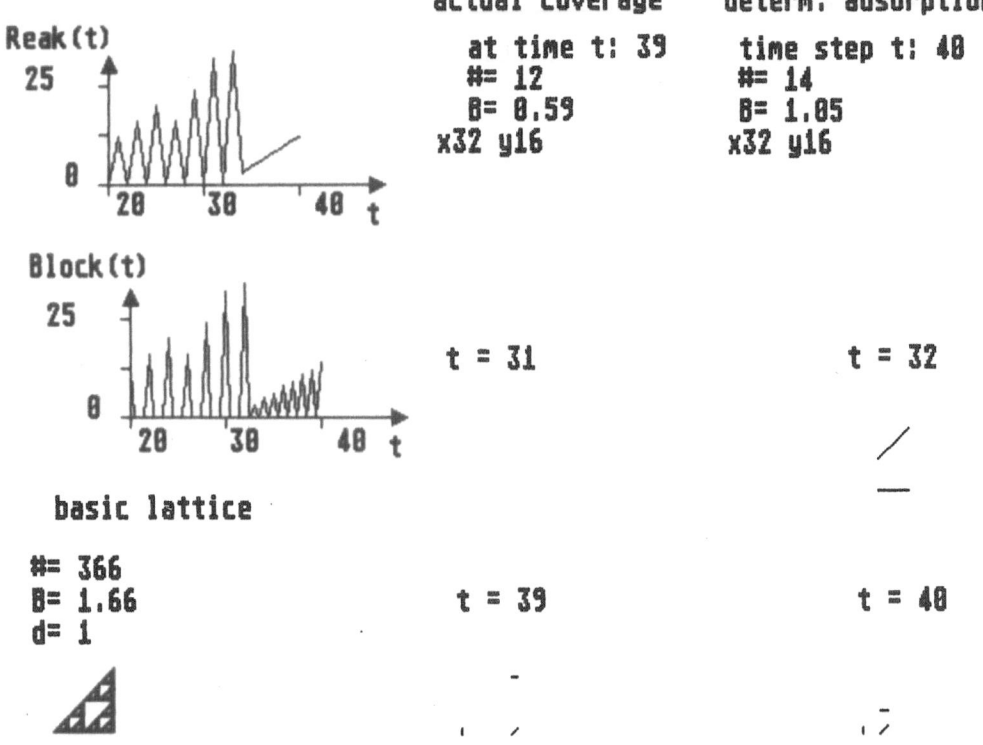

Fig. 19 Spreading of a "circular" wave on a Sierpinsky lattice. The starting cell was c(16,32). The historic rules of the automaton are described above. After reaching the "borderline" of the lattice the two linear sets of cells in state 1 vanish while creating three new starting cells. The development of the coverage [see the function Block(t)] of the lattice with cells in the state 1 oscillates and resembles the development of a one-dimensional 1,2,3,4,5-rule automaton. In this way the one-dimensional automaton is embedded in this fractal one.

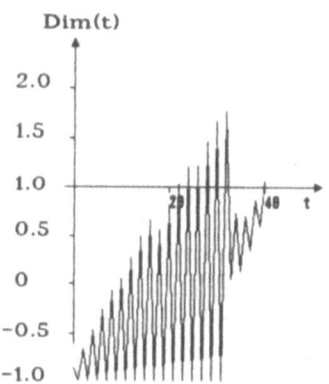

Fig. 20 Oscillation of the growth rate dimension during the development of the automaton shown in Fig. 19.

5. Back to the Experiment

Various automata have been proposed in this article, each of which explains some special aspects of the experimental observations during the catalytic oxidation of CO. I developed these automata in such a way that finally the simulations come close to the chemical reality. Comparing the last rules with the rule previously proposed [23 -25], the temporal recourse to "historic states" seems to be the most interesting feature. In this way an acceleration has been introduced. This acceleration is of the high order of four for the fractal automaton and it is of the usual order of two in the case of the linear automata.

The acceleration only concerns the formation of the non-productive reactors but not their destruction. This last statement is only true if the temperature of the real reactors is high enough at all times t. Otherwise there is a strong negative acceleration (inhibition) for raising the non-productive state as can be seen from the experimental results [24]. For very low CO concentrations in the feed, only small oscillations could be observed [23,24]. This can be explained either by the restriction of the size of the automaton or by the number of coexistent non-productive reactors at time t.

On increasing the CO content, this upper limit of states $z_i(t) = 1$ will also increase. The travelling circular wave will spread over a larger area of the automaton.

For high amounts of CO one may assume that several cyclic waves are created. Those waves annihilate one another and create new starting points so that all in all one can observe a high number of non-productive elementary reactors which may or not oscillate.

6. Conclusion

I have proposed different cellular automata to simulate the heterogeneously catalysed oxidation of CO. Using binary automata the complex behaviour of the chemical reality could only be modelled by complex rules involving large areas of "neighbouring" cells and recourse to "historic" states several time steps back. The reduction to a binary automaton was originally stimulated by the assumption of phase transitions between the two states Pd and PdO during reaction.

It is well known that these states differ widely in their chemical behaviour with respect to the adsorption of CO and O_2. [24,49,69,70]. But I left this microscopic level of description of the reaction and switched to a macroscopic level of productive and non-productive elementary reactors. In this way I gained the freedom to play with the cellular automaton rules without being restricted to the well-founded chemical interpretation.

It was primarily my aim to present methods of qualitative description of the experimentally resulting temporal patterns. To stay within the tradition of discrete chemical description using terms such as atoms and molecules I have chosen the tool of discrete mathematics. Although I only set out to simulate chemical reality by use of cellular automata it turned out that in a way the cellular automata rules could be chemically interpreted. In this restricted way the cellular automata used became models of the reaction. Allow me to emphasize once more that I do not claim to have developed chemical models but I am merely making evident the connections within the observed patterns of the experimental time series. However, if the internal correlations within the pictures to some extent also reflect the dynamics of the real system, I have understood much more about catalysis than I could have expected from playing with such a nice toy as cellular automata.

References

1. E.G. Schlosser, Hetrogene Katalyse, ChT 18, Verlag Chemie, Weinheim (1972).
2. C.N. Hinshelwood, Reaktionskinetik gasförmiger Systeme (translated into German by E. Pietsch and G.Wilcke), Leipzig, akademische Verlagsges. (1928).
3. G.M. Schwab, Katalyse vom Standpunkt der chemischen Kinetik, Springer Verlag, Berlin (1931).
4. D.D. Eley, Advances Catalysis 1, 157 (1948).
5. G. Rienäcker, and J. Völter, Wiss. Fortschr. 23, 403 and 465 (1973).
6. K.F. Bonhoeffer, Z. Elektrochem. u. angew. phys. Chemie 51, 24 (1948).
7. K.F. Bonhoeffer, and G. Langhammer, Z. Elektrochem. u. angew. phys. Chemie 51, 61 (1948).
8. K.F. Bonhoeffer, and G. Vollheim, Z. Naturforschung 8 b, 406 (1953).
9. U.F. Franck, and L. Meunier, Z. Naturforschung 8 b, 396 (1953).
10. U.F. Franck, Angew. Chemie 90, 1 (1978).
11. Hugo, Ber. Bunsenges. Phys. Chemie 74, 121 (1970).
12. H. Beusch, P. Fieguth, and E. Wicke, Chem.-Ing.-Techn. 44, 445 (1972).
13. E. Wicke, H. Beusch, and P. Fieguth, Advances in ACS Symp. Ser. 109, 615 (1972).
14. G. Ertl, P.R. Norton, and J. Rüstig, Phys. Rev. Lett. 49, 177 (1982).
15. P. Möller, K. Wetzl, M. Eiswirth, and G. Ertl, J. Chem. Phys. 85, 5328 (1986).
16. P. Cox, G. Ertl, and R. Imbihl, Phys. Rev. Lett. 54, 1725 (1985).
17. R.J. Schwankner, M. Eiswirth, P. Möller, K. Wetzl, and G. Ertl, J. Chem. Phys. 87, 742 (1987).
18. R. Imbihl, this volume.
19. J.C. Brown, C.A. D'Netto, and R.A. Schmitz, Temporal Order, Springer Series in Synergetics (eds. L. Rensing and N.I, Jaeger, ser. ed. H. Haken), Vol 29, (1985) p.86.
20. L .F. Razon, S.H. Chang, and R. Schmitz, Chem. Engin. Science 41, 1561 (1986).
21 P.C. Pawlicki, and P.A. Schmitz, Chem. Engin. Prog. 40 (1987).
22. J.R. Brown, and R.A. Schmitz, Multiplicity of States in Systems of interacting Catalyst Particles, Preprint August (1988).
23. N.I. Jaeger, K. Möller, and P.J. Plath, Ber. Bunsenges. Phys. Chemie 89, 633 (1985).
24. N.I. Jaeger, K. Möller, and P.J. Plath, J. Chem. Soc., Faraday Trans. I, 82, 3315 (1986).

25. P.J. Plath, K. Möller, and N.I. Jaeger, J. Chem. Soc., Faraday Trans. I., **82**, 1751 (1988).

26. P. Svensson, N.I. Jaeger, and P.J. Plath, J. Phys. Chem. **92**, 1882 (1988).

27. N.I. Jaeger, G. Schulz-Ekloff, and A. Svensson, Proc. 7th Intern. Zeolite Conf., Tokyo 1986, Kadanska Tokyo, and Elsevier Amsterdam, Tokyo (1986) p. 923.

28. A. Kleine, P.L. Ryder, N.I. Jaeger, and G. Schulz-Ekloff, J. Chem. Soc. Faraday Trans. I **82**, 205 (1986).

29. M.M. Slin'ko, N.I. Jaeger, and P. Svensson, J. Catal. in print (1989).

30. A. Dress, N.I. Jaeger, and P.J. Plath, Theoret. Chim. Acta (Berl.) **61**, 437 (1982).

31. A. Dress, M. Gerhardt, N. Jaeger, P. Plath, and H. Schuster, Temporal Order (eds. L. Rensing, and N.I. Jaeger; ser. ed. H. Haken), Springer series in Synergetics, Vol **29**, 67 (1985).

32. M. Gerhardt, H. Schuster, and P.J. Plath, Ber. Bunsenges. Phys. Che. **90**, 1040 (1986).

33. J.C. Mankin, and J.L. Hudson, Am. Inst. Chem. Eng. J. **32**, 1208 (1986).

34. J.C. Mankin, and J.L. Hudson, Chem. Eng. Sci. **41**, 2651 (1986).

35. P.J. Plath, and H. Prüfer, Z. phys. Chem., Leipzig **268**, 235 (1987).

36. P.J. Plath, Proc. Synergetics Meet., Madrid, Oct. 1987, World Scientific Publ. & Co., (1989) in press.

37. N.I. Jaeger, P.J. Plath, and E.v. Raaij, Z. Naturforschung **36a**, 395 (1981).

38. N.I. Jaeger, R Ottensmeyer, and P.J. Plath, Ber. Bunsenges. Phys. Chem. **90**, 1075 (1986).

39. N.I. Jaeger, R. Ottensmeyer, P.J. Plath, and H. Engel-Herbert, Chem. Eng. Sci. (1988) in press.

40. J. Fraissard, T. Ito, L.C. de Manorval, and M.A. Sringuel-Huet, Stud. Surf. Sci. Catal. **12**, 179 (1982).

41. H. Schrübbers, G. Schulz-Ekloff, and G. Wildeboer, Stud. Surf. Sci. Catal. **12**, 261 (1982).

42. J. Sauer, H. Haberlandt, and W. Schirmer, Stud. Surf. Sci. Catal. **1 8**, 313 (1984).

43. V. Kanazirev, V. Penchev, Chr. Minchev, U. Ohlereich, and F. Schmidt, Stud. Surf. Sci. Catal. **18**, 329 (1984).

44. H. Haken, Erfolgsgeheimnisse der Natur - Synergetik, die Lehre vom Zusammenwirken, Deutsche Verlagsanstalt, Stuttgart (1981).

45. A. Wunderlin, and H. Haken, Condensed Matter, Z. Phys. **B 44**, 135 (1981).

46. M. Bestehorn, R. Friedrich, A. Fuchs, H.Haken, A. Kuhn, A. Wunderlin, this volume.

47. A.Th. Haberditzl, N.I. Jaeger, and P.J. Plath, Z. phys. Chem.,Leipzig **265**, 449 (1984).

48. N.I. Jaeger, K. Möller, and P.J. Plath, Z. Naturforschung **36a**, 1012 (1981).

49. K. Möller, Thesis, University Bremen (1984).

50. M. Bülow, Z. Chem. **25**, 81 (1985).

51. J. Caro, M. Bülow, W. Schirmer, J. Kärger, W. Heink, H. Pfeifer, and S.P. Zdanow, J.Chem. Soc. Faraday Trans. I, **81**, 2541 (1985).

52. J. Kärger, and H. Pfeifer, Z. Chem. **16**, 85 (1976).

53. D. Prinz, and L. Riekert, Ber. bunsenges. Phys. Chem. **90**, 413 (1986).

54. G. Bugge, Das Buch der Großen Chemiker, Bd. 2, Verlag Chemie, Weinheim (1974) p. 178.
55. A.F. Holleman, and E. Wiberg, Lehrbuch der anorganischen Chemie, Walter de FGruyter & Co., Berlin (1960) p.15.
56. P.J. Plath, diskrete Physik molekularer Umlagerungen, Teubner Texte zur Physik, Bd. 19, (ser. eds. W. Ebeling, W. Melling, A. Uhlmann, B. Wilhelmi), Teubner, Leipzig (1988).
57. I. Ugi, D. Marquarding, H. Klusascek, G. Gokel, and P. Gillespie, Angew. Chem. **82**, 741 (1970).
58. I. Ugi, Chimia **40**, 340 (1986).
59. P.J. Plath, and J. Schwietering, in preparation.
60. St. Wilson, Physica **10D**, 69 (1984).
61. M. Gerhardt, Thesis, University Bremen (1988).
62. A.K. Dewdney, Spektrum der Wissenschaften, H. **10**, 8 (1988).
63. N. Wiener, A. Rosenblueth, Arch. Inst. Cardiol. Mex. **16**, 205 (1946).
64. T. Toffoli, N. Margolus, Cellular Automata Machines, The MIT Press, Cambridge/Mass., London (1987) p. 47.
65. W. Ebeling, L. Schimansky-Geier, and Ch. Zülicke, this volume.
66. M. Markus ,personal communication on the occurrence of circular waves in two-dimensional cellular automata with a non-quadratic lattice.
67. A. Dress, M. Gerhardt, and H. Schuster, From chemical to biological organization, (edits. M. Markus, S.C. Müller, G. Nicolis), Springer Verlag, Heidelberg, (1988) p.134.
68. J.K. Park, K. Steiglitz, W.P. Thurston, Physica **19D**, 423 (1986).
69. P.V. McKinney, J. Am. Chem. Soc. **54**, 4498 (1932).
70. P.V. McKinney, J. Am. Chem. Soc. **55**, 3626 (1933).

The Study of Kinetic Oscillations in the Catalytic CO-Oxidation on Single Crystal Surfaces

R. Imbihl

Fritz-Haber-Institut der Max-Planck-Gesellschaft, Faradayweg 4-6,
D-1000 Berlin 33, Germany

1. Introduction

In the past decade the study of non-equilibrium systems has attracted a great deal of interest as more and more mathematical tools and theoretical concepts have been developed to try to understand the complex dynamical behavior of these systems. Although the prototype of such a chemical reaction system is still the Belousov-Zabotinskii (BZ)-reaction which takes place in the homogeneous fluid phase, there have also been numerous investigations of kinetic oscillations in heterogeneously catalyzed reactions amongst which the catalytic CO-oxidation has attracted most interest [1,2]. This can be attributed to the simplicity of the chemistry $CO+1/2O_2 \rightarrow CO_2$: However, despite numerous studies, reviewed recently by Schmitz [2] there still exists no commonly accepted mechanism among the different models which have been proposed.

Most of the difficulties in determining the mechanism of kinetic oscillations can be traced back to the lack of experimental data about the actual surface processes under oscillating conditions, as the vast majority of these investigations were carried out under high pressure conditions (p>1 Torr) where very few experimental techniques are available for in situ surface studies. Measurements of the reaction rate alone do not generally allow one to discriminate between various possible reaction mechanisms. These have been performed almost exclusively with the exception of several Fourier Transform Infrared (FTIR) studies [3-8]. The need for more specific information about the nature of the catalyst and the dynamic processes on the surface has also been demonstrated by the results of surface science studies which showed that in addition to structural properties, chemical contaminants can also have a drastic influence on the adsorption properties of a metal. Moreover, it was demonstrated that even the kinetics of simple surface processes can become rather complex due to energetic interaction between the adparticles and the substrate. These interactions may modify the dynamics of the gas/solid interaction and lead to island formation and a phase transition within the chemisorbed layer or even the substrate surface layer. It then becomes a simple proposition to study kinetic oscillations under well-defined conditions in an ultrahigh vacuum (UHV) environment. One can then use the analytical tools of surface science to characterize the surface and follow the changes in both the adsorbate coverages and the surface structure.

The study of kinetic oscillations on Pt single crystal surfaces was started by Ertl's group in Munich [18] where most of the reseach into this subject has taken place to date. The conditions of these investigations, which were generally performed in the 10^{-5} and 10^{-4} Torr range, are significantly different from those of the high pressure studies, as the low pressure ensures strictly isothermal conditions, by extensive cleaning cycles, the formation of Pt oxides could be suppressed. The operation of an oxidation-reduction cycle which has been offered as an explanation for the kinetic oscillations under high pressure conditions [9] could thus be ruled out for the Pt single crystal studies. The results of the investigations summarized in this report show that the mechanisms of kinetic oscillations on various Pt single crystal surfaces can all be explained by one common principle which is that the variations in the reaction rate are due to periodic structural changes in the surface, which modify the catalytic activity. The individual steps of the microscopic mechanism could well be corroborated by in situ studies and numerical simulations could reproduce most of the experimentally observed features. In addition to the mechanistic details, these studies have also revealed a variety of interesting phenomena such as spatial pattern formation and transition to chaos via a series of period-doublings. Two examples, Pt(100) and Pt(110), will be presented here in detail. These have been especially well characterized and represent two limiting cases with respect to their oscillatory behavior. They also represent two limiting cases of spatial self-organization, as we shall see in the next section.

2. Experimental Methods and Reaction Conditions

2.1 Experimental Tools

The experiments [10] were all carried out in a standard all-metal UHV-system of about 100 ℓ volume which was pumped by a 360 ℓ/s turbomolecular pump. Under reaction conditions which were typically in the 10^{-5} and 10^{-4} Torr region, gases were introduced via leak valves while the system was being continuously pumped. The whole chamber was thus operated as a gradient-free flow reactor. The system was equipped with a differentially pumped quadrupole mass spectrometer to measure the reaction rate, with facilities for low energy electron diffraction (LEED), for measurements of the work function via a Kelvin probe and with a cylindrical mirror analyzer (CMA) for Auger electron spectroscopy (AES).

The technique which proved most convenient for observing kinetic oscillations was the continuous measurement of the work function change, due to its high sensitivity and its non-destructive interaction with the surface. The work function of a metal is defined as the energy difference between the Fermi level and the vacuum level and is changed upon adsorption as each adsorbate complex is associated with a non-zero dipole moment. Variations in the work function can therefore be used to monitor changes in the adsorbate coverages. However, since adsorbed oxygen has a much higher dipole moment than adsorbed CO, the work function change in CO/O_2-mixtures mainly reflects the change in the oxygen coverage. Technically, measurements of the work function change were performed using the self-compensating vibrating capacitor method [12] in which an inert ring (diameter ~4 mm) of oxidized Ta wire served as the counter-electrode while the crystal itself represented the other half of the capacitor.

Ordered structures on a single crystal surface can be detected by LEED which due to the strong interaction of electrons with matter, is sensitive mainly to the topmost layer. Additional periodicities on the surface introduced either by an ordered adsorbate overlayer or by lateral displacement of the topmost layer of metal atoms away from their bulklike positions ("reconstruction") can be recognized immediately in the diffraction pattern by the formation of additional spots besides the 1x1 pattern of the substrate lattice. Quantitative measurements of the diffraction pattern were carried out by a computer-controlled video LEED system in which the intensity distribution within a certain selected area of the diffraction pattern could be digitized and stored as data points. In contrast to the work function measurements, LEED probes only a small part of the surface given by the diameter of the LEED beam (~0.5 mm). The local character of LEED measurements can be used for laterally resolving measurements in which the electron beam is deflected magnetically across the surface by means of two pairs of Helmholtz coils. One thus obtains a picture of the intensity distribution of a certain diffraction spot across the surface and if the scan time (typically 2 s to 10 s) is small enough, compared to an oscillation period, this technique can be used to detect and follow spatial pattern formation at various stages of an oscillation cycle, see Section 5.2.

2.2 Sample Preparation and Reaction Conditions

The Pt single crystal samples were cut by spark erosion from a Pt single crystal rod after proper orientation by Laue diffraction. They were rectangular in shape or in the shape of small discs typically 1 mm thick and about 0.5 cm^2 in surface area. They were cleaned in UHV by numerous cycles of oxidation in O_2, followed by sputtering with Ar ions and annealing up to 1000°C. The main impurities C, Ca, P and S could thus be removed very effectively as verified by AES measurements, but the critical test for the sample cleanliness was in fact not AES, but the criterion of whether the sample was no longer oxidizable even in severe O_2-treatments (10^{-4} Torr). The formation of oxides on Pt surfaces has been linked to the presence of Si [13] which does not show up in AES in small concentrations and its removal, as verified by the criterion given above, requires a large number (n>20) of cleaning cycles. Oxides can easily be differentiated from chemisorbed oxygen due to their low reactivity and high thermal stability. Chemisorbed oxygen reacts several orders of magnitude faster with CO than oxides do and it desorbs between 600°C and 900°C, whereas the thermal decomposition of oxides generally requires annealing temperatures above 900°C. The formation of oxides which undoubtedly takes place in the high pressure experiments due to the presence of contaminants [14] can therefore be ruled out under the experimental conditions used here.

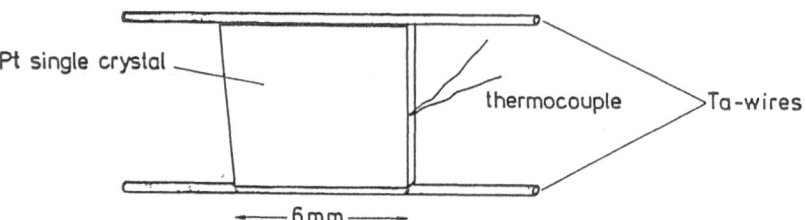

<u>Fig. 2.1:</u> Design used for holding and heating Pt single crystals.

The sample was mounted onto the manipulator via two Ta-wires to which it was spotwelded at the upper and lower edge as indicated in Fig. 2.1. These wires also served to resistively heat the sample by passing current through them, while the sample temperature was measured by a chromel/alumel thermocouple spotwelded to the rear face of the crystal. From the heat released by the surface reaction it can be calculated that temperature changes due to changes in the reaction rate will always be smaller than 0.1 K in the 10^{-4} Torr region, which ensures that the oscillation takes place under isothermal conditions. Gases were introduced by a feedback stabilized gas inlet system which served to establish constant partial pressures as well as to modulate the partial pressures in experiments investigating the response of the system to periodic perturbations, see Section 5.4. Extreme care had to be taken with the gas purity as even small amounts of hydrogen or hydrocarbons were found to completely suppress kinetic oscillations [23].

3.   Conditions and Phenomenology of Kinetic Oscillations

3.1 Instabilities Due to the Langmuir-Hinshelwood Mechanism

It has been well established in numerous studies that the catalytic CO-oxidation takes place via the Langmuir-Hinshelwood (LH) mechanism which proceeds along the following steps [15]:

$$(1) \quad CO + * \quad \rightleftharpoons \quad CO_{ad}$$

$$(2) \quad O_2 + 2* \quad \longrightarrow \quad 2O_{ad}$$

$$(3) \quad O_{ad} + CO_{ad} \quad \longrightarrow \quad CO_2 + 2*$$

* denotes a free adsorption site

Two basic observations are important in the kinetics of the catalytic CO-oxidation. Adsorbed CO can inhibit the adsorption of oxygen and the formation of a

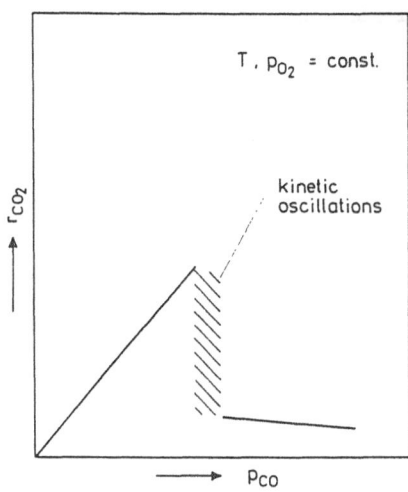

Fig. 3.1: Stable and unstable regions in the kinetics of the catalytic CO-oxidation.

large CO coverage will therefore always be associated with a low reaction rate. In marked contrast, the adsorption of CO is only weakly affected by the presence of adsorbed oxgen. Thus CO can still adsorb and react even on a completely oxygen covered surface. The kinetics of the LH-mechanism are best demonstrated in a diagram in which the reaction rate is plotted versus p_{CO} as shown in Fig. 3.1. In principle, two branches of the reaction rate exist: a high reaction rate branch associated with an oxygen covered surface, in which the reaction rate increases proportional to p_{CO} and a low reaction rate branch associated with a CO-covered surface where the reaction rate decreases with increasing p_{CO}. In the high reaction rate branch, CO molecules impinging on the surface react with a high probability (close to one) and the reaction rate in this branch is therefore determined by the impingement rate of CO molecules e.g. by p_{CO}. If p_{CO} is continually increased, the number of adsorbing CO molecules will finally exceed the number of CO molecules consumed by the reaction and the transition to a CO covered surface will occur on which the rate of oxygen adsorption is rate-limiting. In the transition regime between the two branches, an instability region, indicated by the shaded area in Fig. 3.1 exists, where hysteresis effects and/or kinetic oscillations can be found. Although the non-linearities which exist in the kinetic equation derived from the LH-mechanism, e.g. the quadratic dependence of O_2-adsorption on the number of free sites can predict multiple steady states and hence the hysteresis found in the instability region, these non-linearities are not strong enough to cause kinetic oscillations [2]. These have to be provided by an additional mechanism for which numerous suggestions have been made [2]. In the next section we illustrate how the interplay between the LH-mechanism and a surface phase transition causes kinetic oscillations.

3.2. Review of Investigated Systems

In many cases, the structure of a metal surface is identical to the bulk structure, but in some cases the surface may "reconstruct". The surface atoms then occupy different positions from those in a bulk-like termination. Of the three low-index planes of Pt only the (111) surface is structurally stable, while the (110) and (100) orientations reconstruct as shown in Fig. 3.2. On the clean Pt(100) surface, the Pt-atoms in the surface layer move away from the bulk-like 1x1 positions to form a quasi-hexagonal configuration which will simply be denoted as "hex" [16]. The same quasi-hexagonal configuration, whose high coordination number of six within the surface plane apparently represents a state of minimum free energy, is also obtained by the reconstruction of the (110) surface, but in a different way. Here, every second [110] row is removed, which explains the name "missing row" model for the reconstructed surface such that microfacets with (111) orientations are formed [17].

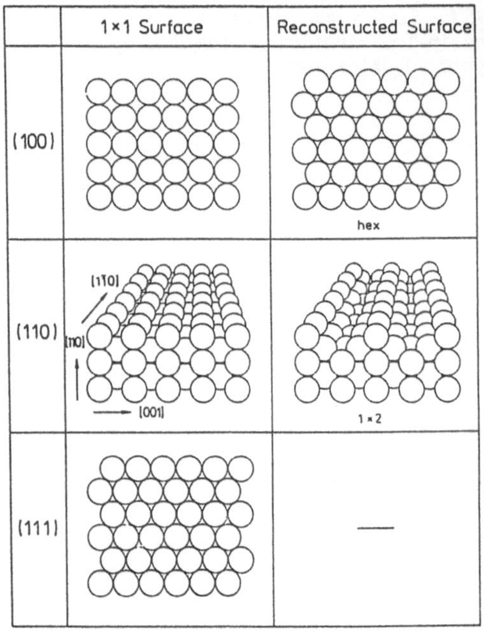

Fig. 3.2: Surface structure for the three low-index planes of Pt in their non-reconstructed and reconstructed state.

It should be added that the different density of atoms in the reconstructed layer requires significant mass transport of Pt atoms, which in the case of the 1x1→hex transition of Pt(100) is 20% of the surface atoms, but amounts to up to 50% in the formation of the 1x2 reconstruction of Pt(110).

Table 1 displays a list of the various single crystal surfaces which have been investigated under low pressure conditions for occurrence of kinetic oscillations in

Table 1

Low pressure studies of kinetic oscillations on single crystal surfaces*

Reaction	Surface	Kinetic Oscillations	p-Range [Torr]	Clean Surface Recon-structed	Suggested Mechanism	References
$CO + 1/2\ O_2 \rightarrow CO_2$	Pt(100)	yes	$10^{-5} - 10^{-3}$	yes	SPT	[14,18-28,45]
	Pt(110)	yes	$10^{-5} - 10^{-3}$	yes	SPT, Facetting	[29-38, 68]
	Pt(111)	no	$< 10^{-3}$	no	-	[14,19,24]
	Pt(210)	yes	$10^{-4} - 10^{-1}$	u.i.	SPT, u.i.	[39]
	Pd(110)	yes	$10^{-3} - 1$	no	Subsurface oxygen	[40,41]
$CO + NO \rightarrow CO_2 + 1/2\ N_2$	Pt(100)	yes	$10^{-9} - 10^{-6}$	yes	SPT, u.i.	[42,43]

SPT=surface phase transition
u.i. = under investigation
*= Kinetic oscillations in the catalytic CO oxidation have also been detected on Pt(100), Pt(111) and Pt(13,1,1) under conditions of a contaminated surface and at atmospheric pressure [14].

catalytic CO-oxidation. The oxidation of CO with O_2 has been studied on all three low-index planes of Pt and remarkably, only the two orientations, Pt(100) and Pt(110), whose clean metal surface is reconstructed, exhibit oscillations, while the structurally stable (111) surface does not show any oscillatory behavior. These observations led to the idea that reversible structural changes of the metal substrate are responsible for the occurrence of kinetic oscillation and this was then confirmed by detailed experiments.

A phase transition of the substrate has also been postulated for Pt(210) which has a rather open and therefore potentially unstable surface structure and where oscillations were also detected [39], but inadequate experimental evidence could be gathered to support a detailed model on this surface. The phase transition 1x1↔hex could presumably also be responsible for the kinetic oscillations in the CO oxidation with NO on Pt(100), as suggested by the authors [42, 43], however, a direct proof by in situ LEED experiments is still lacking. If one tries to extrapolate the idea of a mechanism based on a surface phase transition from Pt to other catalytically active metals, then no oscillations should be found on Pd as clean Pd surfaces do not reconstruct. The fact that recently kinetic oscillations were also detected on Pd(110) [40,41] seems to contradict this, but these oscillations, occurring only at pressures higher than 10^{-3} Torr, were caused by a different mechanism based on the formation of subsurface oxygen, as could be shown by a detailed investigation of the kinetics [41]. Instead of structural changes, the catalytic activity is modulated by chemical changes through the formation of subsurface oxygen. The different mechanisms on Pd and Pt surfaces can be traced back to the different chemical behavior of the two metals as Pd has a stronger tendency to form oxides and easily incorporates oxygen whereas pure Pt has a much weaker tendency to form oxides.

3.3 Kinetic Oscillations on Pt(100) and Pt(110)

Oscillations on Pt(100) and Pt(110) were typically established by first choosing suitable fixed values of p_{O_2} and T and then increasing p_{CO} in a stepwise manner until a sharp decrease of the work function indicated the transition from an oxygen covered surface to a CO-covered surface. Within the p_{CO}-range of this transition characterizing the region of kinetic instability (see Fig. 3.1), spontaneous oscillations can be observed. An example of oscillations established on a Pt(110) surface is displayed in Fig. 3.3 which shows that the reaction rate and the work function both

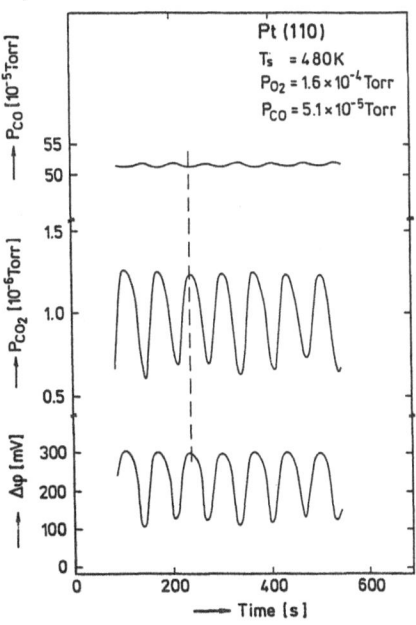

Pt (110)
T_s = 480K
P_{O_2} = 1.6 × 10^{-4} Torr
P_{CO} = 5.1 × 10^{-5} Torr

Fig. 3.3: Variation of the work function (Δφ) with time. The CO_2 partial pressure (p_{CO_2} being proportional to the reaction rate) and of the CO partial pressure during kinetic oscillations on a Pt(110) surface. (After [35]).

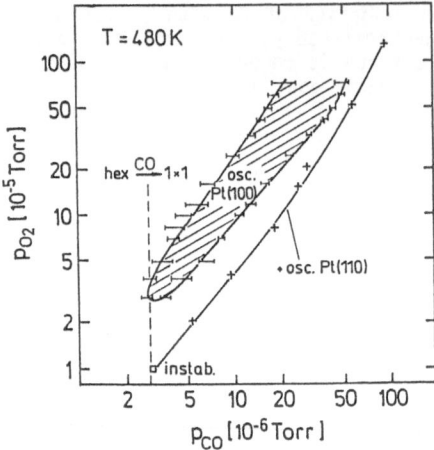

Fig. 3.4: Existence diagram for the occurrence of kinetic oscillations on Pt(100) and Pt(110) (for T=480 K). The narrow existence region for oscillations on Pt(110) is represented by a single line in the diagram. (After [35]).

vary in parallel [34]. This relationship is a consequence of the fact that adsorbed CO inhibits oxygen adsorption and hence the reaction while CO can still adsorb and react onto adsorbed oxygen. The catalytic activity of the surface is therefore determined by the oxygen coverage, whose value is reflected by the work function.

Oscillations on Pt(100) were detected in the 10^{-5} and 10^{-4} Torr range between ~400 K and 550 K with periods typically lasting from around one to several minutes [24]. Their systematic analysis suffered from the fact that the oscillations on Pt(100) in general displayed irregular behavior and only under selected conditions which were difficult to reproduce, could regular oscillations be maintained. Thus no systematic analysis of the dependence of frequency and amplitude on the external parameters was conducted. Despite the irregular behavior of these oscillations, the existence region is highly reproducible and extends over a broad range of the experimental parameters as shown in Fig. 3.4 for T=480 K [23]. The diagram shows that at a given temperature a definite low pressure region exists, below which it is impossible to adjust kinetic oscillations by variation of the ratio p_{O_2}/p_{CO}. The broadness of the oscillation region and the existence of a low pressure limit represent features characteristic of the oscillations on Pt(100) which have to be reproduced by a suitable kinetic model. The boundaries of the oscillation region can be analyzed in terms of kinetic instabilities in the catalytic CO-oxidation on Pt(100) as has been performed by Dath et al. [45].

The oscillatory behavior of Pt(110) represents the opposite of Pt(100) as very regular oscillations can be observed on this surface with a variety of different wave-forms ranging from square-wave-like oscillations to harmonic behavior [29,30,32]. In the pressure range investigated from 10^{-5} Torr to 10^{-3} Torr, oscillations could be established between 310 K and 590 K, with periods ranging from 1 s to 1 h. A reasonable distinction between different oscillatory patterns can be made with the sample temperature. Very regular and fast oscillations with small to medium amplitudes were obtained at high temperatures (T>530 K for $p_{O_2}=1\times10^{-4}$ Torr), whereas slow oscillations (τ>1 min) with large amplitudes which in general exhibited a more irregular character were found at lower temperatures (T<490 K for $p_{O_2}=1\times10^{-4}$ Torr). In the intermediate temperature region, large square-wave like oscillations can be observed on which superimposed fast small amplitude oscillations can be found. This type of oscillation apparently represents a transitional state between oscillations of the high temperature type and those of the low temperature type. The existence region of kinetic oscillations on Pt(110) which is also displayed in Fig. 3.4 differs markedly from those of Pt(100) as it is rather narrow and extends only over ~2% of p_{CO} at low temperatures and up to 10% at higher temperatures [29].

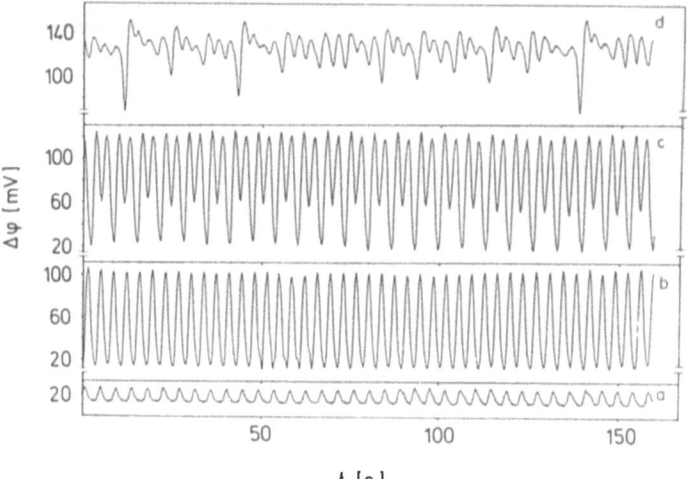

Fig. 3.5: Development of kinetic oscillations and transition to chaotic behavior as p_{CO} is decreased in small steps starting from the level of the CO-covered surface with p_{CO}=4.2x10^{-5} Torr (a) to 3.9x10^{-5} Torr (d). The temperature T=530 K and p_{O_2}=8x10^{-5} Torr were kept constant. (After [34]).

chaotic behavior via a series of period-doublings which appear to follow the well-known Feigenbaum scenario that has been observed in many other fields of physics [29,34]. This transition, which is depicted in Fig. 3.5, occurs as the control parameter p_{CO} is decreased going from regular oscillations, which occur near the CO-covered surface of the low reaction rate branch, (see Fig. 3.1) to chaotic oscillations which are found in the vicinity of the reaction rate maximum. The deterministic nature of the chaotic oscillations in Fig. 3.5 could be proven by detailed mathematical analysis [34] in which the various criteria discriminating against random behavior (noise) such as positive Lyapunov exponents, a positive and finite Kolmogorov entropy, and convergence of the embedding dimension were shown to be fulfilled. The embedding dimension which is the minimum number of variables needed to describe the system, was determined from the calculation of the correlation integrals to be between 5 and 6, as can be seen from the scaling region (the region between the dashed lines) in Fig.3.6 where the slope of the curves calculated for different embedding dimensions remains approximately constant beyond n=5.

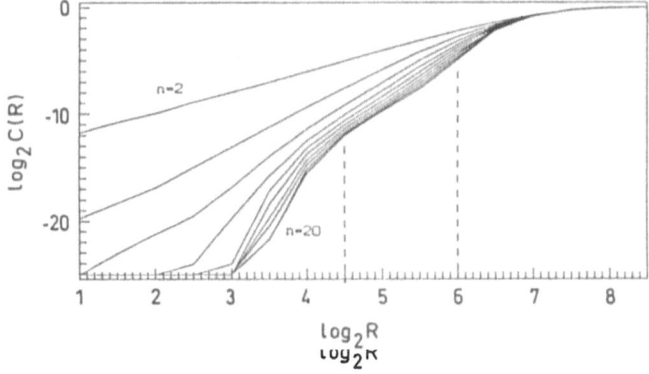

Fig. 3.6: Correlation integral C(R) for various embedding dimensions n calculated for an aperiodic (chaotic) time series as shown in Fig. 3.5. (After [34]).

A particularly interesting phenomenon which can be observed in the high temperature oscillations on Pt(110) is the transition from regular oscillations to

4. Mechanism of the Low Pressure Oscillations

4.1 Mechanism of the Adsorbate-Induced Phase Transition

The removal of the surface reconstruction by an adsorbate gives rise to a reversible surface phase transition as a change in the adsorbate coverage causes the surface to switch between the reconstructed and the non-reconstructed configuration. The switching of the stability depending on the adsorbate coverages can be explained by simple thermodynamics summarized in the energy diagram displayed in Fig. 4.1 and which were first discussed for Pt(100)/CO [44].

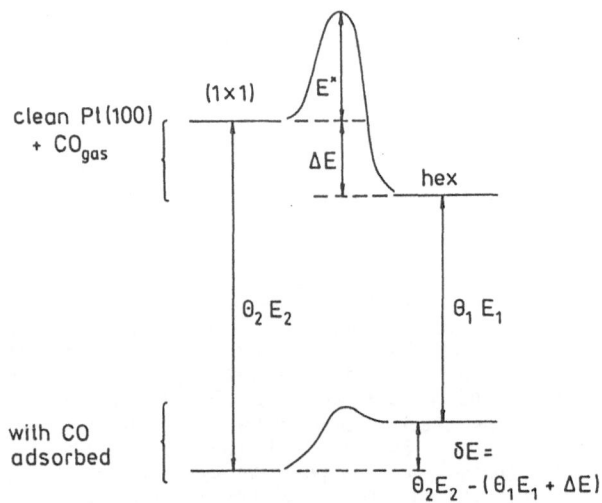

Fig. 4.1: Energy diagram for the CO-induced removal of the hex reconstruction on Pt(100). E_1: adsorption energy of CO on the hex phase (27 Kcal/mol); E_2: adsorption energy of CO on the 1x1 phase (35 Kcal/mol for $\Theta_2=0.5$); ΔE: energetic stabilization of the clean hex phase. (After [44]).

The clean surface reconstructs because the reconstructed surface is lower in its free energy than the bulk-terminated 1x1 surface. If an adsorbate is put on top of the surface then its adsorption energy has to be added to the surface free energy in order to obtain the energy of the system surface+adsorbate. This will cause a switching of the stability of the two phases if the adsorption energy is substantially higher on the 1x1 than on the reconstructed phase as is indicated for the system Pt(100)+CO in Fig. 4.1. The reconstruction will in that case be lifted by the adsorbate and as the reconstruction energy of the clean surface has to be overcompensated by the gain in adsorption energy, it is apparent that the removal of the reconstruction is controlled by a critical adsorbate coverage. For the system Pt(100)/CO the heat of adsorption is 37 Kcal/mol on the 1x1 phase, but only 27 Kcal/mol on the hex phase and the critical CO coverage for the transition 1x1→hex has been determined to be ~0.25 to 0.30. These values were derived from the hysteresis in the adsorption/desorption equilibrium of CO on Pt(100) that is associated with the phase transition 1x1↔hex [44]. It should be noted that not every adsorbate causes a lifting of the reconstruction, which is plausible from the energetic considerations outlined above. For example, both CO and oxygen remove the hex reconstruction of Pt(100), but only CO lifts the 1x2 reconstruction of Pt(110) which persists with adsorbed oxygen.

4.2. The Oscillatory Mechanism on Pt(100)

4.2.1 Experimental Results

The idea behind a mechanism for kinetic oscillations based on a surface phase transition is that the phase transition modulates the catalytic acitvity of the surface through the different adsorption properties associated with each of the two phases. As the initial sticking coefficient (the probability that a molecule which hits the surface sticks) for oxygen chemisorption on the unreconstructed Pt(100) surface with $s_0 \sim 0.1$ is more than two orders of magnitude higher than on the hex surface, where s_0 is only 10^{-3} to 10^{-4} [46], this idea seemed to be very plausible as an explanation for the oscillations found on Pt(100) and the actual experimental verification was conducted as described [20].

As the two surface phases are associated with the different diffraction patterns displayed in Fig. 4.2a, one would expect an in situ LEED experiment to show the

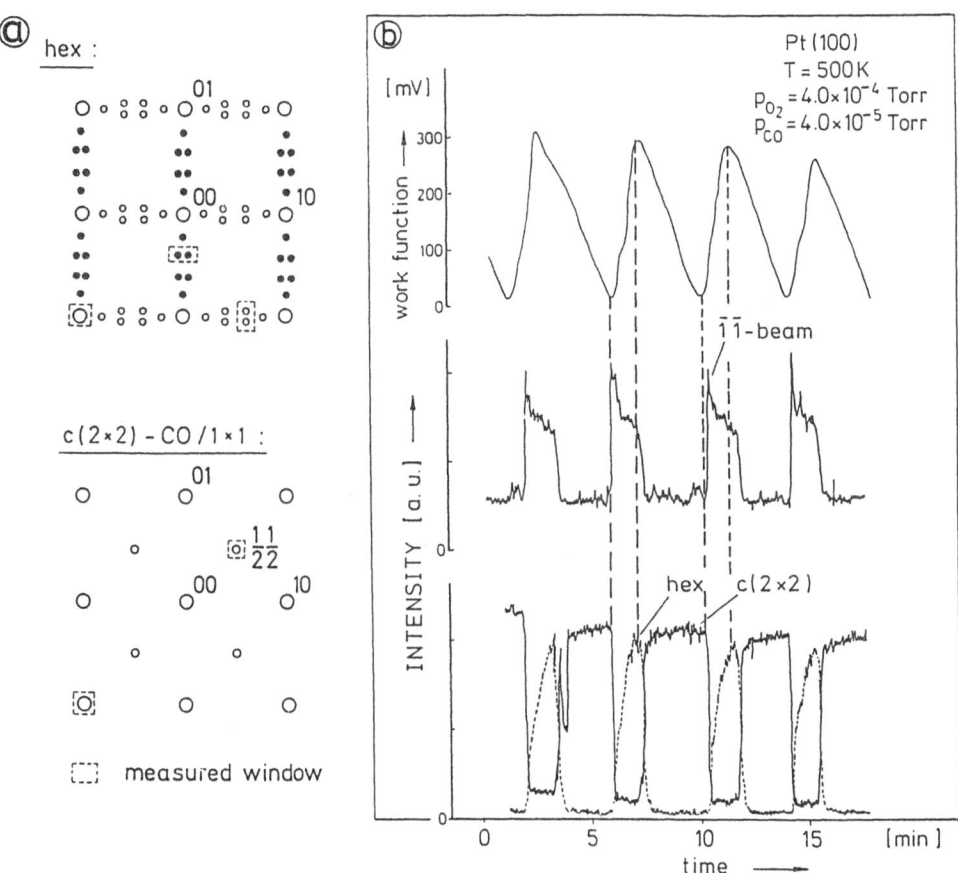

Fig. 4.2a: Schematic sketches of the LEED patterns for the reconstructed and unreconstructed Pt(100) surface. CO adsorbed on the 1x1 phase forms an ordered c(2x2)-overlayer with $\Theta_{CO}=0.5$. Broken lines enclose specific diffraction spots whose intensity was recorded as a function of time during kinetic oscillations, (see Fig. 4.2b).

Fig. 4.2b: Periodic structural changes of the Pt(100) surface during kinetic oscillations. (After [20]).

structural changes of the surface through the changes in the diffraction pattern. A periodic switching of the surface between the hex and the 1x1 configuration which accompanies the oscillations in the reaction rate was in fact observed, as demonstrated by the oscillations in the LEED intensities and work function in Fig. 4.2b. The latter quantity served as a measure for the reaction rate. Since the work function could not be measured simultaneously with LEED, its phase relationship to the LEED intensities in Fig. 4.2b are best considered arbitrary. Since CO forms an ordered overlayer, the plateau of nearly constant c(2x2) intensity just reflects a CO covered 1x1 surface which is then rather abruptly transformed into the hex phase. Unlike CO, oxygen does not produce an ordered overlayer under the oscillation conditions and the question was one of how the large oxygen concentration, which could be deduced from the magnitude of the $\Delta\phi$-amplitude, was distributed on the surface. However, the presence of oxygen showed up indirectly in LEED through the pronounced intensity increase of the 1̄1-beam in between the disappearance of the c(2x2) pattern and the intensity maximum of the hex phase [24]. The analysis of the experiments, in terms of structural and adsorbate coverage changes, showed that a CO-covered c(2x2)/1x1-surface with a CO-coverage around 0.5 is replaced by an oxygen covered 1x1 surface which is then transformed into a nearly adsorbate free hex phase and the hex phase is finally converted back into the CO-covered c(2x2)/1x1-surface.

4.2.2 Mechanism and Numerical Simulations

For a kinetic model we need to describe the variations in the adsorbate coverages as well as in the surface structure. We start with the kinetic equations of the LH-mechanism [2] and let the adsorption properties e.g. the corresponding constants depend on the substrate configuration. The substrate configuration is changed by the phase transition 1x1↔hex which is controlled by critical adsorbate coverages [21]. The adsorbate free 1x1 surface will reconstruct, while the adsorption of CO causes the removal of the hex reconstruction. We denote the fractions of the surface present in the 1x1 and hex configuration by a and b with a+b=1. The corresponding oxygen and CO coverages on the two phases are termed accordingly as v_a and v_b, u_a and u_b respectively. As the oxygen sticking coefficient is very small on the hex phase we can immediately set v_b to zero. We thus arrive at a set of four coupled differential equations that describe the variations in the CO coverage on the 1x1 (1) and on the hex phase (2), the oxygen coverage on the 1x1 phase (3) and the phase transition of the substrate (4) which is governed by critical adsorbate coverages [21].

$$\frac{du_a}{dt} = k_1 a p_{CO} - k_2 u_a + k_3 a u_b - k_4 u_a v_a / a \tag{1}$$

$$\frac{du_b}{dt} = k_1 b p_{CO} - k_6 u_b - k_3 a u_b \tag{2}$$

$$\frac{dv_a}{dt} = k_7 a p_{O_2} \left\{ \left[1 - \frac{2u_a}{a} - \frac{5}{3} \frac{v_a}{a} \right]^2 \right.$$

$$\left. + \alpha \left[1 - \frac{5}{3} \frac{v_a}{a} \right]^2 \right\} - k_4 u_a v_a / a, \tag{3}$$

$$\frac{da}{dt} = \begin{cases} \dfrac{1}{U_{a,grow}} \dfrac{du_a}{dt}, & \text{if } U_a > U_{a,grow} \text{ and } \dfrac{du_a}{dt} > 0 \quad (4a) \\[3ex] -k_8 a(1-c), & \text{if } c = \dfrac{U_a}{U_{a,crit}} + \dfrac{V_a}{V_{a,crit}} < 1 \quad (4b) \\[3ex] 0, & \text{otherwise.} \quad (4c) \end{cases}$$

CO adsorption and desorption on the 1x1 and hex phase is treated in the terms containing k_1, k_2 and k_6. For oxygen adsorption we need two adjacent empty sites which are given by the first term following k_7. As we know from the experimental results [52] that a c(2x2)-CO adlayer with $\Theta_{CO}=0.5$ does not completely inhibit oxygen adsorption as has been assumed in the first term, we add a second term with the prefactor α where CO does not block adsorption sites and which accounts for the existence of defects in the CO adlayer. The introduction of the "defect" term was very important for the simulation, as no oscillations could be obtained without its existence. The surface reaction between CO_{ad} and O_{ad} described by the term containing k_4 has to proceed entirely on the 1x1 phase.

An additional term with K_3 is introduced in equations (1) and (2) which takes into account the unidirectional diffusion of CO adsorbed on the hex phase to the 1x1 area where it is "trapped" due to its higher adsorption energy on the 1x1 phase. This term is important for the phase transition hex→1x1 treated in (4a), since we assume that the additional CO in the 1x1 islands provided by the "trapping" term causes these islands to grow, thus maintaining a constant local coverage $U_{a,grow}$, which is near 0.5 [44,67]. The phase transition in the direction 1x1→hex, which is described by (4b) takes place if the adsorbate coverage falls below a critical limit necessary to stabilize the 1x1 phase. This transition is strongly activated by 20 to 25 Kcal/mol [43] in contrast to the phase transition hex $\xrightarrow{CO}$ 1x1 which even proceeds at 150 K [44].

Although nearly all of the constants used in (1) to (4) could be taken from experimental values published in the literature, the main emphasis in the simulation was put on the reproduction of qualitative features, as a meaningful quantitative analysis would have required a much more elaborate analysis of the model [21]. An initial test of the model was carried out with the simulation of the hysteresis which occurs in the adsorption/desorption equlibrium of CO on Pt(100). This hysteresis, associated with the phase transition 1x1↔hex can be ascribed to the existence of nucleation barriers in the formation of the new phase as is commonly found in first order phase transitions. The good agreement between experimental and simulated data reproduced in Fig. 4.3 confirms that the treatment of the adsorbate induced phase transition in the model was correct. Using the experimental parameters of oscillatory conditions, the model produces kinetic oscillations whose qualitative features agree very well with the experimental findings, as can be seen by a comparison of the simulated oscillations in Fig. 4.4 with the experimental data in Fig. 4.2b. The model underlying the kinetic equations (1) to (4) will be described in the next paragraph using the simulated oscillations in Fig. 4.4.

We adjust the experimental parameters ($p_{O_2}/p_{CO} \sim 10{:}1$) with a large excess of oxygen such that an oxygen covered 1x1 surface would result as the stable state, if the 1x1 configuration were frozen in.

(1) For the starting point, we choose a CO-covered 1x1 surface which could be maintained as a metastable state for an infinite amount of time if the inhibition of oxygen adsorption worked perfectly. However, as the CO-adlayer contains

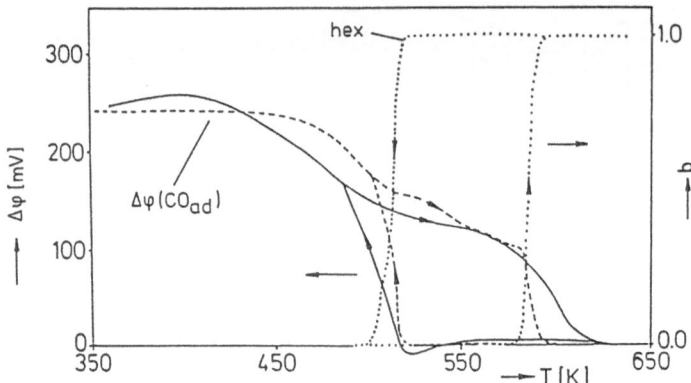

Fig. 4.3: Comparison of the measured (full line) and calculated (dashed line) hysteresis in the adsorption of CO on Pt(100). A Pt(100) surface is held in a CO atmosphere with p_{CO} kept constant, while the temperature is varied continuously in a heating/cooling cycle. The work function $\Delta\phi$ serves as a measure for the CO coverage u_a. The dotted lines monitor the hex reconstruction b. (After [21]).

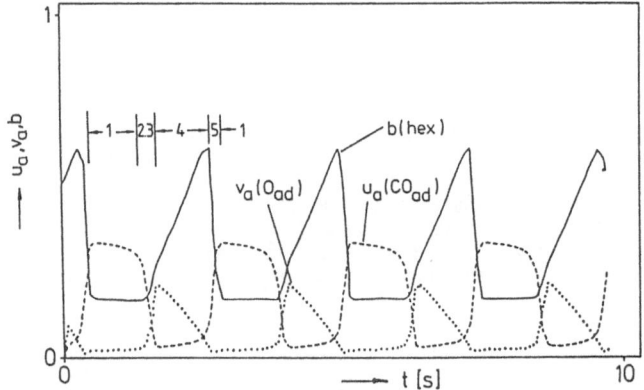

Fig. 4.4: Calculated variation in the adsorbate coverages (u_a=CO, v_a=oxygen) and of the hex reconstruction (b) during kinetic oscillations on Pt(100). In the experimental curves of Fig. 4.2b, u_a corresponds to the c(2x2)-intensity, v_b to the 11-beam intensity and b to the hex intensity. (After ref. 21).

    defects where oxygen can still adsorb, we observe a slowly decreasing plateau of CO concentration.

(2) Below a certain level of CO concentration, the reaction accelerates very rapidly and the CO-adlayer is completely reacted away in a very short time. This is due to the autocatalytic kinetics of this step as the surface reaction creates empty sites where oxygen can adsorb and reactively remove the residual CO.

(3) A nearly adsorbate free 1x1 surface is formed as a reactive intermediary. Due to the high oxygen sticking coefficient of the empty 1x1 surface, oxygen adsorbs readily. Here, it is important to note that the clean metastable 1x1 phase is separated by a large activation barrier from the stable hex phase (see Fig. 4.1) which gives oxygen the opportunity to adsorb as the reactive 1x1 phase would otherwise immediately be converted into an unreactive hex phase. Some hex phase will nevertheless be formed, but the larger part of the surface will be transformed into an oxygen covered surface.

(4) Without the phase transition, the oxygen covered 1x1 surface would now be the stable state of the surface due to the experimental parameters initially chosen e.g. a large excess of oxygen. This surface, however, is destablized by the phase transition in a slow dynamic process. Part of the empty 1x1 sites, created by the ongoing surface reaction, is always converted into the hex phase instead of being filled up by oxygen atoms again. As the hex sites are lost for oxygen adsorption, the oxygen covered 1x1 surface is finally completely converted into the oxygen free hex phase.

(5) A prerequisite for kinetic oscillations on Pt(100) is that the p_{CO} must be high enough to trigger the transition hex $\overset{CO}{\rightarrow}$ 1x1-c(2x2). Since oxygen does not adsorb on the hex phase CO could adsorb onto the clean hex phase and immediately convert it into the c(2x2)/1x1-phase. However this does not happen as long as a substantial amount of oxygen covered 1x1 surface still exists. Since the adsorption energy is higher on the 1x1 phase, CO on the hex phase migrates to 1x1-O islands where it reacts, thus keeping the CO-coverage on the hex phase below the critical coverage necessary to trigger the hex→1x1 phase transition. Only when the oxygen coverage has decreased to a low enough value does the nucleation of c(2x2)-CO islands take place, which leads back to the initial situation.

Numerical simulations of the oscillations carried out with varying parameters p_{O_2}, p_{CO} for T=480 K successfully reproduced the qualitative features of the experimental existence diagram (Fig. 3.4), in particular, the existence of a low pressure limit. The existence of this limit follows directly from the mechanism depicted above. As the phase transition hex→1x1 proceeds entirely under the influence of adsorbed CO, the p_{CO},T-conditions which mark the transition hex→1x1 in the CO-hysteresis, (Fig. 4.3) will set a definite limit in the parameter space of the oscillations as indicated in Fig. 3.4. The treatment of the phase transition in (4) has been improved by Andradé [27] towards a closed description through the use of a Landau-Ginzburg potential derived from coverage dependent interaction energies. The same author also carried out a bifurcation analysis in which he found both Hopf and SNIPER bifurcations at the borderline of the existence region. A kinetic model based on a surface phase transition has also been applied to kinetic oscillations on supported Pt catalysts by Lynch et al. [66].

4.3 The Oscillatory Mechanism on Pt(110)

4.3.1 Observed Structural Changes

The phase transition 1x1↔1x2 of the Pt(110) surface proceeds in a similar way to the phase transition of Pt(100), as the reconstruction is lifted by adsorbed CO. The same mechanism as applied to the kinetic oscillations on Pt(100) could also be expected to operate on Pt(110) as the two surface phases, 1x1 and 1x2, are presumably associated with different adsorption properties. This picture was confirmed by in situ LEED studies which showed that the variations in the reaction rate are in fact accompanied by periodic structural changes [29,32].

A fundamental difference to Pt(100) exists however, as the structural changes on Pt(100) are restricted to the first layer, whereas they can go several layers deep on Pt(110) and eventually cause the appearance of new surface orientations in a facetting process. Thus two kinds of structural transformations exist on Pt(110), superimposed on each other, namely the 1x1↔1x2 phase transition and a facetting process. It will be shown below that the picture does not become too complicated as facetting only plays a role in the low temperature oscillating region [31,32].We can therefore distinguish between high temperature oscillations (T>~500 K) which take place on a flat (110) surface and low temperature oscillations where the surface is facetted (T<~500 K).The connection between kinetic oscillations and facetting is simplified by the fact that facetting can also be observed outside the oscillatory region. The facetting of Pt(110) must therefore be a phenomenon independent of kinetic oscillations (leaving out speculations about possible microscopic mechanisms for facetting). Its strong influence on the catalytic activity and the adsorption behavior will be discussed first. However, under properly selected conditions facetting occurs during kinetic oscillations and in this case the oscillatory behavior of Pt(110) is strongly modified by the facetting process, as shown below.

4.3.2 Conditions for Facetting and Its Influence on the Catalytic Activity

The facetting of the Pt(110) surface which takes place under the influence of the catalytic CO oxidation (but not with one of the gases alone!) can be observed in LEED as a continuous process which first leads to a broadening and then to a splitting of the integral order beams along the [110]-direction, as demonstrated by a series of LEED profile measurements, displayed in Fig. 4.5a [31]. The analysis of the diffraction pattern reveals that preferentially, facets are formed belonging to the [001]-zone where the structural elements consist of steps with (100)-orientation and (110) terraces. The structural elements can easily be identified in a ball model of the (210) facet depicted in Fig. 4.5b, where one (110) terrace unit cell is always followed by a (100) step unit cell. By varying the number of (110) terrace unit cells per (100) step unit cell, the angle of inclination of the facet with respect to the flat (110) plane can be increased/decreased, which then corresponds to larger/smaller splitting of the integral order beams in LEED. The degree of facetting, as judged by the width of the observed beam splitting, depends greatly on the applied reaction conditions e.g. p_{O_2}, p_{CO}, T. The limiting case for strong facetting is represented by the (210) plane in which one (110) terrace unit cell and one (100) step unit cell alternate. LEED only provides limited information about the size of the facets, but

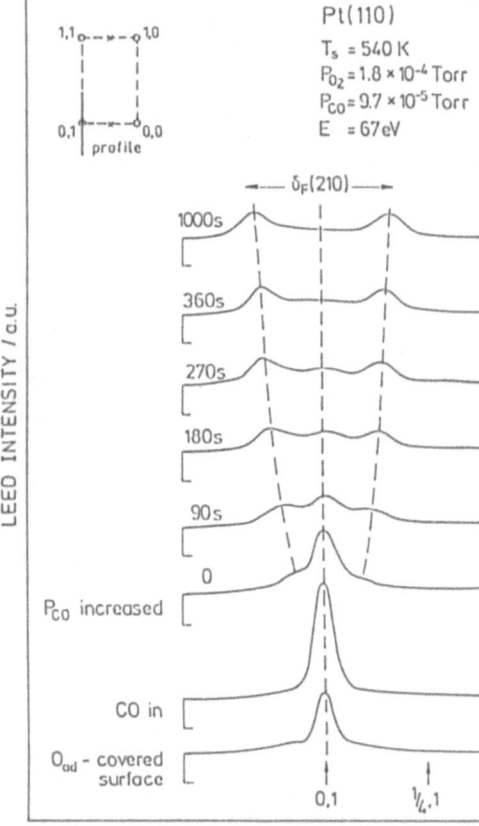

Pt(110)

$T_s = 540$ K
$P_{O_2} = 1.8 \times 10^{-4}$ Torr
$P_{CO} = 9.7 \times 10^{-5}$ Torr
$E = 67$ eV

LEED INTENSITY / a.u.

RECIPROCAL SPACE

Fig. 4.5b: Ball model showing the flat unreconstructed Pt(110) surface in the front and a (210) facet in the rear section.

Fig. 4.5a: Development of (210) facets during catalytic CO-oxidation on Pt(110) as measured through LEED intensity profiles along the [110]-direction. (After [31]).

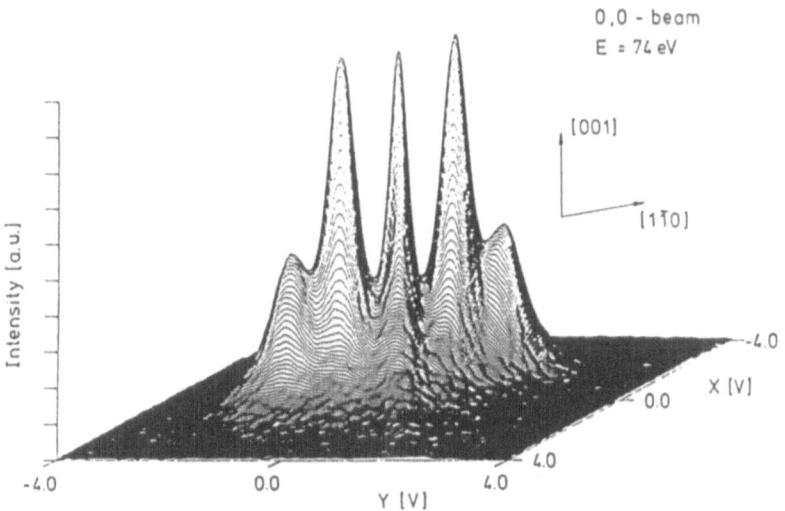

0,0 - beam

E = 74 eV

[001]

[1̄10]

Fig. 4.6: Splitting of the 0,0-beam into five distinct beams as a result of the formation of (430) facets of uniform size on Pt(110). The observed splitting (1.35% of the Brillouin zone) can be explained by an equally spaced (n~70 lattice units) saw-tooth-like array of (430) facets. (After [68]).

from the sharpness of the LEED spots it can be concluded that these are in the order of 10 to 100 Å and include a large amount of structural disorder. These small facets which we can call microfacets in order to distinguish them from facets of macroscopic size observed in many catalytic reactions [48] are thermally unstable and can easily be removed by heating the sample up to ~700 K.

The conditions for facetting and the changes in the catalytic activity caused by facetting can both be best rationalized in terms of a diagram showing the functional dependence of the reaction rate on p_{CO} for fixed values of p_{O_2} and T. The solid line in the lowest section of Fig.4.7 demonstrates this dependence for the unfacetted Pt(110) surface which displays the behavior usually found in catalytic CO-oxidation, see Section 3.1. Facetting was found to occur only beyond the maximum of the reaction rate towards higher p_{CO}, but in a region where the reaction rate is still finite. In the same region, the phase transition 1x1↔1x2 also takes place as the transition from the high reaction rate branch to the low reaction rate branch is associated with a change in the surface structure from an oxygen covered 1x2 surface to a CO covered 1x1 surface. If these observations are translated into facetting conditions, it follows that a CO covered 1x1 surface and a high reaction rate both favor the growth of facets.

The facetting process introduces a new aspect to the understanding of reaction kinetics in the catalytic CO-oxidation as the measured rate curves are usually thought of as being associated with a certain fixed geometry of the catalyst surface. This assumption is definitely wrong in the case of the facetting of Pt(110) and the consequences for the reaction kinetics can be observed directly if one adjusts the conditions with a fixed p_{CO} value beyond the rate maximum such that the surface becomes facetted. Thus the reaction rate does not remain stationary, but slowly evolves towards the high reaction rate branch as indicated by the arrows in Fig. 4.7. By simultaneous LEED measurements, it can be shown that the rate increase is actually due to facetting, as demonstrated in Fig. 4.8 by a parallel increase in the reaction rate and the splitting of the 0,1-beam. The increase in the catalytic activity of Pt(110) resulting from facetting, shows up in a plot of the reaction rate versus p_{CO} as the rate maximum is now shifted towards a higher p_{CO} value. This effect is clearly visible in the rate curves of a strongly and less strongly facetted Pt(110) surface, depicted in Fig. 4.7. An explanation for the increase in catalytic activity was

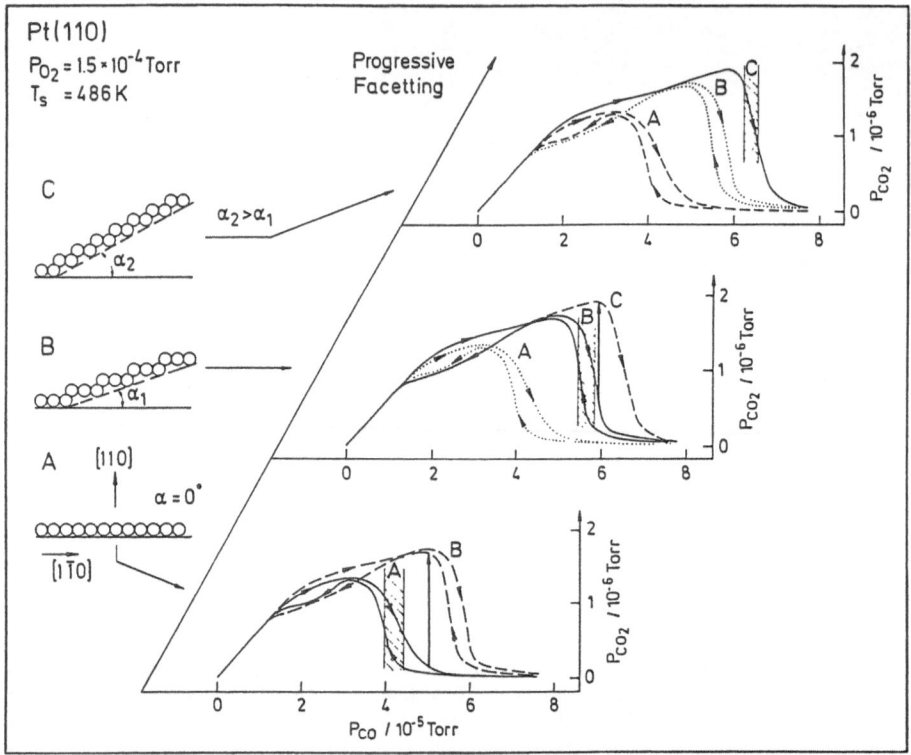

Fig. 4.7: Change in the catalytic activity of a Pt(110) surface as a result of facetting. The rate curves for a flat (110) surface and two surfaces with different degrees of facetting are shown (rate curve C belongs to a surface with (210) facets). The arrows between the rate curves indicate the facetting process, observed as a slow increase of the reaction rate where p_{CO} is kept constant. The shaded areas demonstrate the existence region of oscillations for the three different surface conditions. (After [32]).

offered by subsequent adsorption experiments which showed that the initial sticking coefficient for oxygen adsorption increased strongly (up to a factor of 1.5 to 2) due to facetting [38]. The increase in the oxygen sticking coefficient is due to the lower work function at the steps. This effect facilitates the donation of metal electrons into the antibinding $2\pi^*$-orbital of the O_2-molecule which is a mechanistic step in the dissociation of the molecule. As the density of steps reaches a maximum on the (210) facet, it follows that this surface has the fastest oxygen adsorption kinetics. The increase in the oxygen sticking coefficient has the same effect as increasing p_{O_2} and it therefore becomes clear why the rate maximum is shifted towards a higher p_{CO} on a facetted surface.

The mechanism leading to the formation of facets can be explained by the large mass transport of 50% of the surface atoms associated with the phase transition 1x1↔1x2. As every second row has to be removed from a flat 1x1 surface in order to create the 1x2 "missing row" structure, these atoms have to be moved into a new layer on top of the original layer. Therefore, steps are created representing the initial stage of facetting. As one can easily demonstrate with a ball model, facets can be created by repetition of this sequence, but the essential point is still the lack of a driving force for the facetting of Pt(110). An increase in the adsorption energy on the facets, compared to the flat Pt(110) plane could stabilize the facets and provide a simple thermodynamical explanation for the tendency of Pt(110) to facet in a CO/O_2

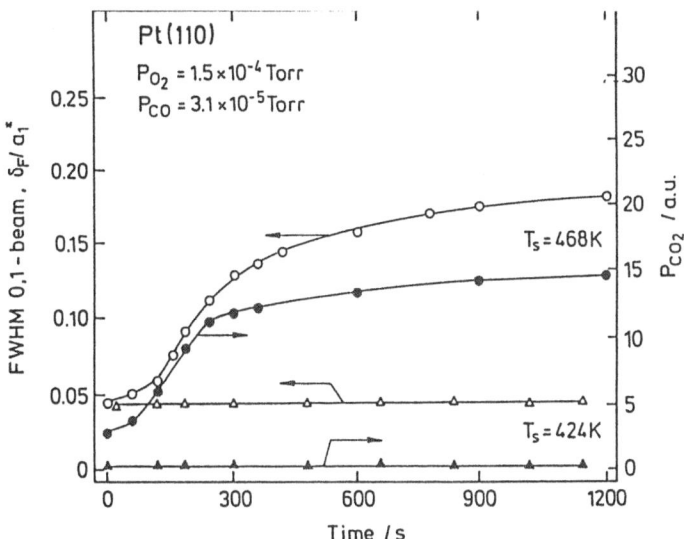

Fig. 4.8: Variation of the steady-state rate of CO oxidation on Pt(110) (outside the oscillatory region) by continuous facetting of the surface as monitored by the LEED beam splitting δ_F. At $T_S=424$ K the rate (full triangles) is practically zero and δ_F (open triangles) stays constant, while at $T_S=468$ K both the rate and the degree of facetting continuously increase. (After [31]).

atmosphere. Such an energetic stabilization has not been found as yet and might in fact be irrelevant as facetting only occurs under typical non-equilibrium conditions which are in general controlled by the kinetics of the dynamic processes involved and to a much lesser extent by the principles of equilibrium thermodynamics. The coverage dependence of the phase transition and the change of the adsorption properties by facetting introduces strong coupling between the surface reaction and the substrate changes where complex feedback behavior itself might direct the structural changes of the surface towards the formation of facets. A cellular automaton simulation of facetting based on the ideas described above, is currently in progress and the first results look rather encouraging. The experimental observation that facetting was always found to be associated with an increase in the reaction rate is probably very important and has so far remained unexplained. This observation clearly demonstrates that the structural changes do not take place randomly, but have to be guided by some ruling principle which still remains to be formulated.

4.3.3 Influence of Facetting on Kinetic Oscillations

As could be anticipated from the results of the preceding section facetting also has a very strong influence on the oscillatory behavior of Pt(110) [31]. This influence is directly visible in the reaction rate traces of Fig. 4.7 as, together with the reaction rate maxima, the oscillation region indicated by the shaded area in the diagram is shifted towards higher p_{CO} values as the surface becomes progressively facetted. This effect seems to complicate the description of the oscillations somewhat as the state of the surface now appears as another parameter in addition to p_{O_2}, p_{CO}, and T. The actual situation, however, is much simpler, since facetting does not develop under conditions of high temperature oscillations and even at a lower temperature, the surface only facets to a certain extent which then remains constant throughout the oscillations (however it may vary within one cycle as will be shown below). Therefore, even the low temperature oscillations will exhibit a high degree of reproducibilitiy as the same degree of facetting will be reached with identical parameters and starting conditions.

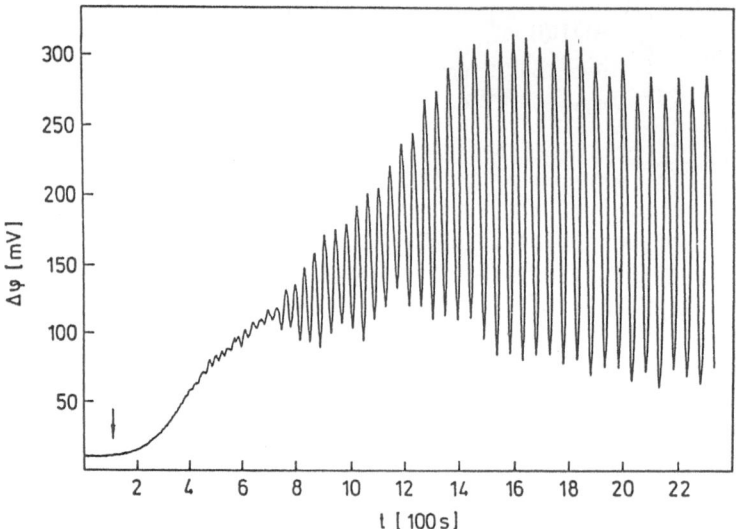

Fig. 4.9: Development of kinetic oscillations by facetting of the Pt(110) surface. T=470 K and pCO=2.3×10^{-5} Torr were kept constant, while the O_2-pressure was changed at the point marked by an arrow from 1.5 to 2.0×10^{-4} Torr and then kept constant again. (After [29]).

One negative consequence of the shift in the oscillatory region due to facetting is that initially, only unstable oscillations are obtained through the period of progressive facetting which calls for frequent readjustments of the experimental parameters in order to remain in the oscillatory region. But facetting also has very nice characteristics for the experimentalist trying to adjust the kinetic oscillations, in that there is an induction period in which the surface automatically evolves towards optimum oscillatory conditions, demonstrated by the gradual development of regular large-amplitude oscillations as displayed in Fig. 4.9. The oscillatory region in this experiment was initially outside the experimentally chosen parameters, but was shifted towards these parameters by facetting which caused the development of oscillations. Besides a shift in the oscillatory region, facetting also leads to a pronounced increase in the length of the oscillation period which in some cases can go up by two orders of magnitude. A typical experiment is shown in Fig. 4.10 where oscillations with a period of 11s were adjusted on an intially unfacetted surface, but within one hour the surface had become facetted causing the oscillation period to increase from 11s to 55s.

As the oscillatory behavior and the catalytic activity of Pt(110) is profoundly changed by the facetting process, one should suspect that facetting is also involved in the oscillatory mechanism itself. This assumption was verified by in situ LEED measurements which showed that the degree of facetting varies periodically in parallel to the oscillations of the reaction rate [31]. Oscillations were adjusted on a strongly facetted surface and LEED profile measurements the results of which are displayed in Fig. 4.11, revealed that the oscillations in the 1x2 intensity and hence the reaction rate are accompanied by periodic changes in the splitting of the integral order beams which serves as a measure for the degree of facetting. This result is quite surprising as it requires substantial mass transport of Pt atoms over distances of the order of the size of the facets and not just lateral displacements of the surface Pt atoms as in the case of the phase transition 1x1↔hex on Pt(100). The changes in the facet orientation and facet size deduced from the LEED results are depicted in Fig. 4.12 for the two extrema of one oscillation cycle. The reaction rate maximum corresponds to the steepest inclination angle of the facets in agreement with the previous result (Fig. 4.7) where the catalytic activity increases approximately

44

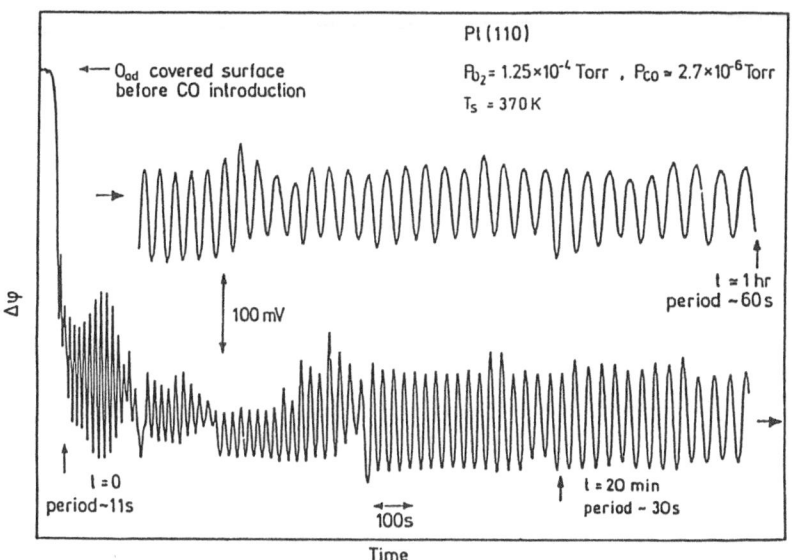

Fig. 4.10: Gradual increase of the oscillation period during the development of facetting. (After [32]).

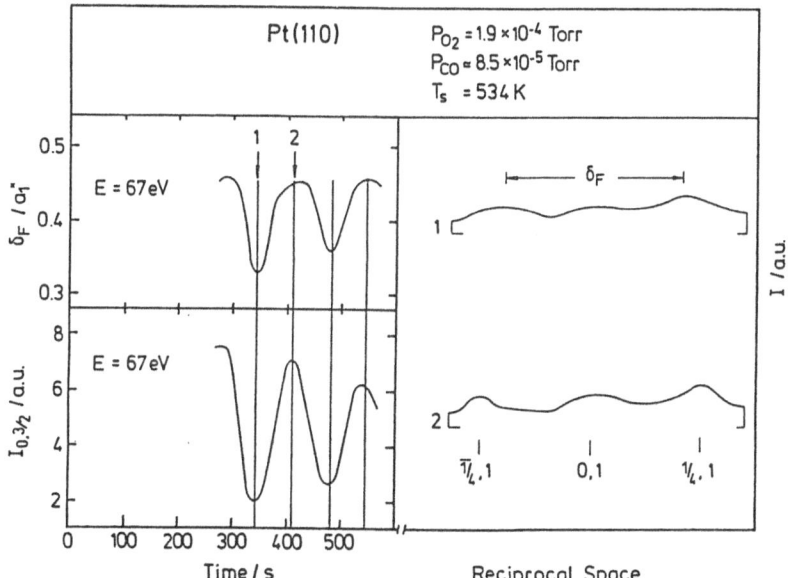

Fig. 4.11: Periodic changes in the degree of facetting during kinetic oscillations. The 0,1-beam splitting δ_F serves as a measure for the degree of facetting, while the reaction rate has been shown to be proportional to the intensity of the 0,3/2-beam of the 1x2 reconstruction. (After [32]).

proportionally to the step density and hence the steepness of the facets. Note that even on the strongly facetted surface, flat (110) areas have to exist, as can be deduced from the observation of the 1x2 reconstruction whose periodic changes cause the oscillations of the 1x2 intensity in Fig. 4.11.

low reaction rate

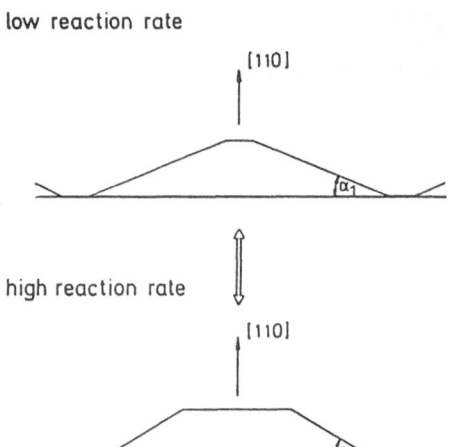

Fig. 4.12.: The facetted Pt(110) surface showing how the changes in the degree of facetting are linked to changes in the size of flat (110) terraces during kinetic oscillations. (After [32]).

high reaction rate

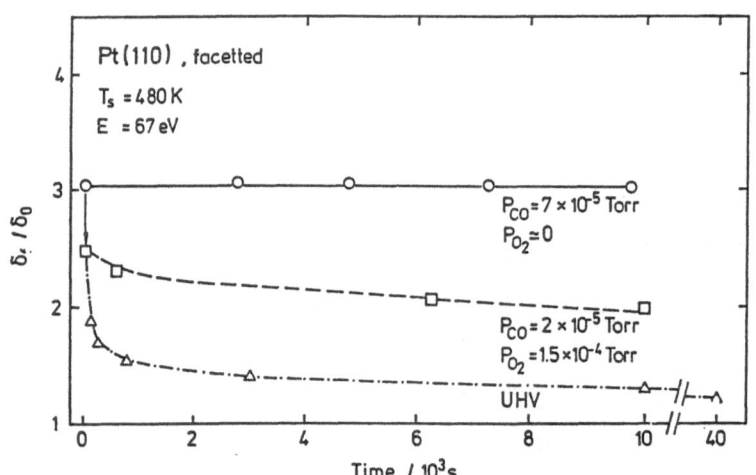

Fig. 4.13: Slow removal of facets by thermal annealing under various reaction conditions. The 0,1-beam splitting δ_f served as a measure for the degree of facetting. (After [32]).

One important part in the description of oscillations on a facetted Pt(110) surface is still missing; the removal of facets in order to render the facetting reversible. This process has been identified as the thermal reordering of the facetted Pt(110) surface towards the flat (110) plane which is the thermodynamically more stable surface due to its lower surface area and possibly its lower surface free energy. It has already been mentioned that the facets can be removed simply by heating the sample, but the process of facet removal also takes place at moderate temperatures under the conditions of the oscillations. This has been demonstrated by an experiment, where the results are shown in Fig. 4.13 in which the facet removal under different reaction conditions at T=480 K was followed by measurements of the 0,1-beam splitting over several hours. The results show that the removal of the facets strongly depends on the surface conditions as the facets were removed gradually if the surface was bare or if reaction conditions corresponding to the high reaction rate branch of the LH-mechanism were adjusted, but the facets were stabilized if the surface was kept completely CO-covered. For the oscillation mechanism, it is important to note that first, both facet growth and removal take place on a time-scale comparable to those of the low temperature oscillations and second, that both processes depend on the adsorbate coverages and hence the reaction conditions.

The idea of a dynamic equilibrium between the growth and the removal of facets offers a simple explanation for the existence of a high temperature limit for facetting, separating the low temperature oscillations which take place on a facetted surface from the high temperature oscillations which occur on a flat (110) surface. As the growth rate for facets is roughly proportional to the reaction rate, which however, is limited by partial pressures, a temperature increase can only lead to a small enhancement of this process, in contrast to the removal of facets, which is limited by the mobility of the surface Pt atoms and therefore can increase very sharply as the temperature is raised. As a consequence, the balance between facet growth and facet removal is shifted with an increase temperature towards the latter and above a certain temperature limit, the surface remains essentially flat.

4.3.4 Discussion of the Mechanism

For a discussion of the mechanism involved, we will first focus on the simpler case of the high temperature oscillations which take place on a flat Pt(110) surface. The only structural change we observe in LEED is the 1x1↔1x2 phase transition and, provided that the two phases exhibited different adsorption properties, the description of the mechanism would proceed completely analogous to Pt(100). The latter assumption could not be determined unequivocally, although the experimental evidence points toward a higher sticking coefficient for oxygen on the 1x1 than on the 1x2 phase [47]. The difference will in any case be small compared to Pt(100) as the reconstructed phase already exhibits a high sticking coefficient with $s_0 \sim 0.3$ to 0.4. In computer simulations it could be shown that the small difference in s_0 for the two phases has very important consequences for the oscillatory behavior, because the existence region for oscillations narrows down in p_{O2}/p_{CO} parameter space as the difference in the adsorption behavior of the two phases becomes smaller [35]. The strikingly different width of the existence region for oscillations on Pt(100) and Pt(110) (Fig. 3.4) is the result of the different magnitude of the changes caused by the phase transition.

During the low temperature oscillations we observe both changes in the degree of facetting _and_ the 1x1↔1x2 phase transition, but for the discussion of the mechanism we consider only the effects caused by a varying degree of facetting, as these were shown to have the main influence on the catalytic activity of the surface. We use a simplified representation of the reaction rate curves for two different degrees of facetting (Fig. 4.7) which is displayed in Fig. 4.14 and discuss the conditions which determine the growth and removal of facets. If we start with a weakly facetted surface covered by CO (point A), then we have conditions which favor the growth of facets while the opposite process, thermal reordering, will be ineffective because the facets are stabilized by adsorbed CO. Due to the growth of facets, the sticking coefficient for oxygen will increase and consequently the reaction rate will slowly rise towards the rate maximum (point B). At point B, the surface conditions will have changed and instead of a CO-covered 1x1 surface we will observe an oxygen covered 1x2 surface. This change in the surface conditions will make the facet growth rate very low, while the thermal reordering process, favored by these conditions, becomes dominant. Consequently we will observe the removal of facetting as the net effect and the reaction rate will decrease until point A is reached again, thus completing one cycle.

The mechanism presented here for kinetic oscillations on Pt(110) contains one unpleasant feature, as two different mechanisms, albeit separated by temperature, have been assumed to operate on the same single crystal surface. The distinction between a mechanism based on the phase transition and a mechanism based on reversible facetting, disappears however, if one considers both mechanisms on an atomic scale in the limit of a weakly facetted surface. This follows simply from the properties of

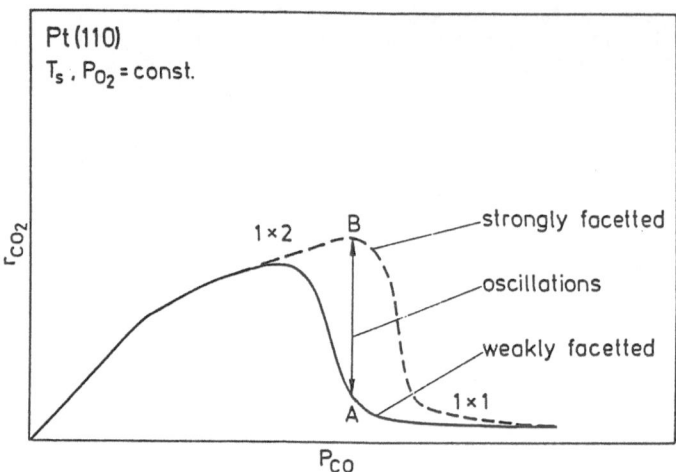

Fig. 4.14: Schematic representation of the rate curves of Fig. 4.7 for a weakly facetted and a strongly facetted surface. Oscillations are assumed to occur via periodic changes in the degree of facetting, corresponding to transitions between A and B. (After [32]).

the phase transition 1x1↔1x2 which involves extensive mass transport and therefore creates steps which may be regarded as the initial stage of facetting. Small changes in the degree of facetting will also be present in the high temperature oscillations. This does not contradict the LEED results which gave the impression of a constantly flat surface through the oscillations, as such small changes in the degree of facetting will not show up in LEED, due to its limited instrumental resolution. As both mechanisms become identical in the limit of a weakly facetted surface, the apparent difference between them has to be quantitative and not qualitative in nature. To be more specific, the dependence on the temperature shows that it is in principle an ordering effect which in one case leads to the formation of facets and in the other case leaves the flat surface intact but with a high density of steps and defects.

As a change in the degree of facetting requires massive mass transport of Pt atoms, one could speculate that the strong increase in the oscillation period which is observed as the surface becomes progressively facetted, is due to kinetic barriers in the mass transport of the Pt atoms which increase as the facetting becomes stronger. A more conclusive answer might be provided by a kinetic model for oscillations on Pt(110) which, unlike that for Pt(100), is not yet available.

4.4 Oscillations Via Formation of Subsurface Oxygen

The study of kinetic oscillations on Pd surfaces enables an interesting comparison to be drawn between the behavior of the Pd and Pt surfaces, as both metals have similar catalytic properties, but differ in their structural surface properties and in their tendency to form oxides. At temperatures below 440 K and for $p>10^{-3}$ Torr, the Pd(110) surface exhibits kinetic oscillation during the catalytic CO oxidation whose existence region in p_{O_2}, p_{CO}, T-parameter space has been mapped out between 10^{-3} Torr and 1 Torr [40]. As clean Pd surfaces do not reconstruct, a mechanism similar to that presented above for clean Pt surfaces seemed to be unlikely, but an explanation for the driving force of the oscillations on Pd(110) was soon provided when the formation of subsurface oxygen was detected [41]. This species, located in the near-surface region of bulk Pd is known to alter the catalytic activity of the surface [49] and a mechanism based on the slow formation/removal of subsurface oxygen was in fact suggested several years ago by Turner, Sales and Maple [9] to explain kinetic oscillations found at high pressure on polycrystalline Pt. The effect of subsurface oxygen also shows up clearly in the hysteresis of the CO_2 production rate on Pd(110) which develops characteristic features in the oscillating region which are not present on Pt surfaces nor on Pd(110) outside the oscillating region e.g. at higher temperatures and/or lower pressures [41].

The hysteresis in the reaction rate on Pd(110), complemented by simultaneous measurements of the work function change is shown in Fig. 4.15. The catalytic CO-oxidation on Pd(110), in fact exhibits two different hysteresis loops which only overlap under oscillating conditions. Unlike Pt(100) and Pt(110), the highest work function increase on Pd(110) is caused by adsorbed CO and not by adsorbed oxygen and the rise of the work function in Fig. 4.15 reflects the transition from an oxygen covered to a CO covered surface. As can be seen, for example, by comparison with the measurements on Pt(110) in Fig. 4.7, the behavior of the hysteresis in Fig. 4.15 is significantly different from that derived simply from the LH-mechanism. However all the characteristic features in the hysteresis obtained on Pd(110) can be consistently explained if we assume that the catalytic activity of the surface is reduced by the presence of subsurface oxygen.

Depending on whether the conditions favor the formation or the removal of subsurface oxygen, we can divide the diagram into two sections which are separated approximately by the rate maximum. In region I at low p_{CO}, the formation of subsurface oxygen takes place as oxygen from the surface penetrates into the bulk region. Obviously this process is facilitated by a large oxygen coverage. The reverse process, the removal of subsurface oxygen, proceeds at high p_{CO} beyond the rate maximum where oxygen from the bulk diffuses to the surface where it reacts with adsorbed CO and leaves the surface as CO_2. This process favored by a large CO

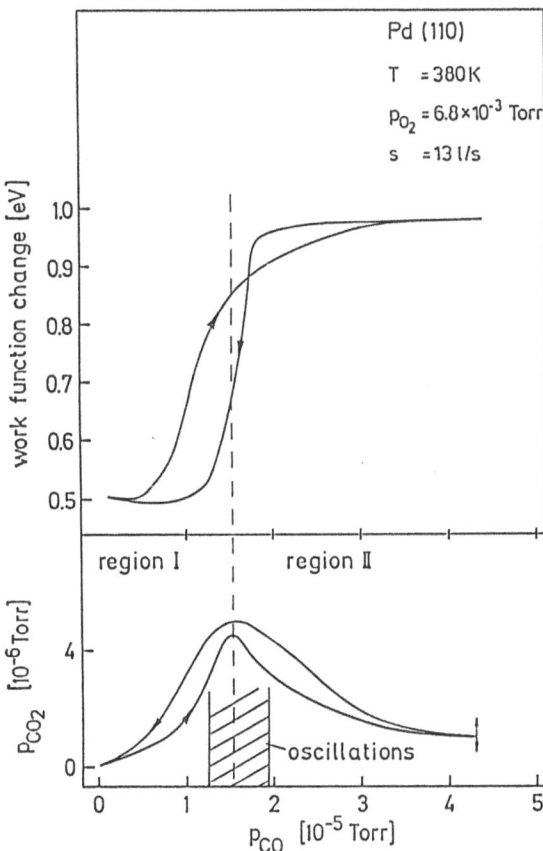

Fig. 4.15: Steady-state measurements of catalytic CO-oxidation on Pd(110) under conditions where oscillations can be found (shaded region). The work-function change and the CO_2 production rate were monitored while p_{CO} was slowly increased and decreased with p_{O_2}, T being kept constant. (After [41]).

coverage. In complete contrast to the behavior of the Pt surfaces, where only one clockwise hysteresis loop is observed centered around the reaction rate maximum, we observe two hysteresis loops on Pd(110) and the one at high p_{CO}, corresponding to the hysteresis loop on the Pt surfaces, rotates in the opposite sense. If we first focus on the hysteresis at low p_{CO} which has no counterpart on the Pt surfaces, then we can attribute the rate multiplicity to the different amount of subsurface oxygen formed depending on whether the same p_{CO} value is reached coming from high p_{CO} or from low p_{CO}. In the first case, less subsurface oxygen will have been formed, due to the lower oxygen coverage at higher p_{CO} and consequently, the activity of the surface and hence the reaction rate will be higher than in the second case where the reverse arguments hold. The most striking difference with respect to the simple LH-mechanism occurs in the second hysteresis, where the catalytic activity of the surface is enhanced on returning from conditions which establish a CO-covered surface, while the reaction rate remains low in the LH-mechanism down to some p_{CO} value where the jump to the stable high reaction rate branch occurs. The origin of the second hysteresis loop on Pd(110) can be explained in the same way as the first hysteresis loop, as again, the amount of subsurface oxygen present depends on the direction in which the p_{CO} has been varied. Therefore a rise in the p_{CO} value which establishes a CO covered surface, leads to a decrease in the amount of subsurface oxygen which in turn leads to an increase in the catalytic activity as p_{CO} is decreased once more.

Kinetic oscillations on Pd(110) were found to occur in the vicinity of the rate maximum and only under conditions where the hysteresis in the reaction rate exhibits the peculiar behavior discussed above [40,41]. We can therefore assume that the oscillation mechanism is directly related to the two hysteresis loops caused by subsurface oxygen and the simplest way to explain the kinetic oscillations is by assuming that these are due to periodic variations in the amount of subsurface oxygen modulating the catalytic activity of the surface. As in the mechanism presented for Pt(100) and Pt(110) we then have a high reaction rate and a low reaction rate branch of the LH-mechanism corresponding to an oxygen and a CO covered surface and the transitions between them are caused by the formation and removal of subsurface oxygen. The active oyxgen covered surface will slowly be deactivated by the formation of subsurface oxygen and as CO is no longer consumed by the reaction, a CO covered surface will result, in turn reactivating the surface by removing subsurface oxygen. This mechanism which follows the scheme suggested by Turner et al. [9] also gives a reasonable explanation for why relatively high pressures, compared to Pt surfaces, are required to obtain kinetic oscillations on Pd(110). The diffusion of oxygen from the bulk to the surface for a CO-covered surface and from the surface into the bulk region for an oxygen covered surface requires the existence of a chemical equilibrium between surface and bulk oxygen in pure O_2 and a minimum p_{O_2} is probably necessary to obtain sufficient bulk concentrations just for thermodynamic reasons.

Since the solution of oxygen in bulk Pd is a precursor state for Pd oxide formation, the chemical nature of Pd subsurface oxygen could resemble Pd oxide. This, however, cannot be the case as subsurface oxygen which diffuses to the surface has a reactivity towards CO comparable to that of chemisorbed oxygen, whereas Pt oxide reacts with CO several orders of magnitude slower than does chemisorbed oxygen [49]. Nor did AES measurements carried out after oscillation experiments show any detectable oxide formation. Therefore the subsurface oxygen observed here with Pd(110) should be considered as a precursor for the formation of a real oxide. It has a different nature than the subsurface oxygen discussed by Turner et al. who attributed to it oxide-like features e.g. a very low reactivity. But as is clearly demonstrated by the hysteresis scheme in Fig. 4.15, subsurface oxygen causes a lowering of the reactivity of the surface which mechanistically can be traced back to several effects. If subsurface oxygen lowers the adsorption energy of CO, then this will decrease the residence time of adsorbed CO and hence its chance to react which is likewise the dominating effect in the first hysteresis where CO adsorption is rate-limiting. In the second hysteresis, beyond the rate maximum oxygen adsorption is rate-limiting and here subsurface oxygen might lower the catalytic activity by reducing the oxygen sticking coefficient or just by decreasing the reactivity of chemisorbed oxygen towards CO. The different influences are difficult to separate and a number of details in the mechanism still remain to be resolved.

4.5 Comparison of the Different Mechanisms

All oscillating systems discussed so far exhibit a common feature in parameter space which is that the oscillations always occur near the rate maximum in the transition regime between the high reaction rate branch and the low reaction rate branch of the LH-mechanism, see Fig. 3.1. In this instability region, the reaction is sensitive enough to small changes in the parameters and in the catalytic properties so that together with the necessary feedback behavior, oscillations may develop.

In the instability region, the sticking coefficient for oxygen becomes rate-limiting for the reaction and the structural dependence of this parameter forms the basis for the oscillatory mechanism discussed for Pt(100) and Pt(110). In both cases, the driving force for the oscillations is the reversible structural changes of the surface which modify the catalytic activity of the surface through the sticking coefficient for oxygen adsorption. What is different is the extent of the structural changes which, in the phase transition 1x1↔hex on Pt(100), involve only the topmost layer of Pt atoms, while the structural changes of Pt(110) may go several layers deep and lead to the formation of facets. These differences can be rationalized by regarding the mass transport which, in the phase transition 1x1↔1x2 of Pt(110), involves 50% of the surface atoms, while only about 20% must be moved in the reconstruction of the Pt(100) surface. The necessary feedback behavior is provided by the coverage dependence of the structural changes which constitutes an interdependence of all three factors; adsorbate coverages, surface structure and reaction rate in the oscillatory mechanism. A prerequisite for obtaining kinetic oscillations on clean Pt single crystal surfaces is the existence of a surface phase transition or a structural instability. This explains why no oscillations were detected on the stable Pt(111) surface.

In a recently published experiment, it was demonstrated that the structural changes of a catalyst under oscillating conditions can even reach almost macroscopic dimensions. After a Pt wire had been exposed to a CO/O$_2$-atmosphere at atmospheric pressure for several hours, facets of the size of several microns could be detected with an electron microscope [50]. This example clearly shows that structural changes of a catalyst also occur under high pressure conditions and it establishes a link between the low pressure single crystal studies and the high pressure experiments with polycrystalline material.

A clear connection exists between the low pressure kinetic oscillations on Pd(110), which have been ascribed to the formation of subsurface oxygen, and the high pressure oscillations on polycrystalline Pt and Pd, where an oxidation/reduction mechanism has been proposed [9]. Although the oxide model offers a plausible mechanism for the high pressure oscillations, the failure to detect the proposed changes in the oxide concentration during kinetic oscillations [6,8,14] has led to the formulation of a new model, the so-called carbon model, in which the active sites of the surface are blocked by a periodically varying carbon coverage [6]. The results obtained on Pd(110) could thus be considered as an experimental confirmation of the oxide model which is still lacking. Although the subsurface oxygen detected on Pd(110) is of a different nature than an oxide, the use of higher oxygen pressure will lead to conversion of the subsurface oxygen species into a real oxide, whose presence might affect the catalytic activity in a similar way as the subsurface oxygen on Pd(110). The comparison between the Pt and Pd surfaces demonstrates that structural, as well as chemical, changes have to be taken into account regarding the mechanism of kinetic oscillations in the catalytic CO-oxidation. Which effects become dominant depends on the conditions applied and on the substrate used.

5. Spatial Self-Organization and Coupling between Oscillators

5.1 Necessity for Spatial Self-Organization

The necessity for a synchronization mechanism between the different parts of a catalyst surface becomes apparent evident from a "Gedankenexperiment" in which one divides the surface area into a number of small, separate compartments which oscillate independently. With a random phase relationship between the different oscillators, their amplitudes would add up to zero or at most cause small fluctuations. Thus a mechanism must exist which synchronizes the oscillations in various parts of the surface in order to obtain measurable macroscopic variations in the reaction rate. More realistically, one would even have to assign different adsorption properties to some or all of the compartments to account for the spatial inhomogeneities of the surface as even a very well prepared single crystal surface exhibits a large number of defects such as steps, dislocations, etc. Again a synchronisation mechanism must exist which is strong enough to enforce regular oscillations over all these structural imperfections in order to explain the nice, regular behavior which is so often observed. One can go a step further and ask to what extent the oscillatory behavior is determined by defects, i.e. acknowledge the possibility that certain structural irregularities might play the role of pacemaker for the rest of the surface.

These considerations clearly illustrate the importance of spatial self-organization on a catalyst surface. As will be shown below, the macroscopic oscillatory behavior e.g. amplitude, regularity and reproducibility is to a large extent determined by differences in spatial self-organization. For communication between different parts of the surface three possibilities exist in principle: energy transfer via thermal coupling, mass transfer via surface diffusion and coupling via concentration changes in the gas phase. The first mode has been shown to be dominant in high pressure oscillations where temperature excursions of more than 150 K have been reported [3,4,7,51]. Here, synchronization is achieved as initially small temperature increases in one part, trigger the reaction in adjacent parts to form reaction zones travelling through the whole reactor [32]. In the single crystal studies quoted this mode has to be discarded however, as the conversion rates are too low and the thermal conductivity of the crystals too high to produce a temperature change high enough to influence the surface reaction.

Each of the latter possibilities; surface diffusion and coupling via the gas phase, has been found in single crystal studies on Pt(100) and Pt(110), but with each surface being dominated by a different mode [35]. The two single crystal surfaces, therefore, represent two limiting cases for spatial self-organization which in the following are shown to be responsible for the differences in their oscillatory behavior, namely the regularity of the oscillations on Pt(110) which is contrasted with the irregular oscillations prevailing on Pt(100). The different type of spatial self-organization on the two surfaces, however, does not represent an isolated property of these systems, but can be traced back to the different adsorption characteristics of the two surfaces, which is also expressed in the different width of the existence region for oscillations, see Fig. 3.4.

5.2 Coupling via Surface Diffusion of Adsorbed CO

Spatially resolving measurements during kinetic oscillations are possible using the Scanning LEED technique in which the single crystal surface is rastered by the primary LEED electron beam while the intensity of the diffraction beams is simultaneously recorded as a function of time. A series of frames is shown in Fig. 5.1, each representing the intensity distribution of the hex and c(2x2) pattern across the Pt(100) surface at a particular moment at various stages of an oscillation cycle [22,24]. The c(2x2) intensity may be taken as a measure of the CO concentration while the hex intensity indicates the amount of reconstruction. The sequence of the frames demonstrates that CO is not uniformly removed across the surface, but instead, one observes a reaction front in Fig. 5.1 propagating from the top to the bottom of the surface, which removes adsorbed CO. At some distance behind the initial reaction front the CO level starts to rise again which then completes the oscillating cycle as the initial situation, with an almost completely CO covered surface, is re-established. One notices that the distribution of the c(2x2) and hex intensity complement each other resulting from the fact that the reconstruction is lifted by adsorbed CO. Comparing the spatially resolved measurements of kinetic oscillations on Pt(100) in Fig. 5.1 with the results of local LEED measurements displayed in Fig. 4.2, one recognizes, as expected, the same sequence of structural changes, but the

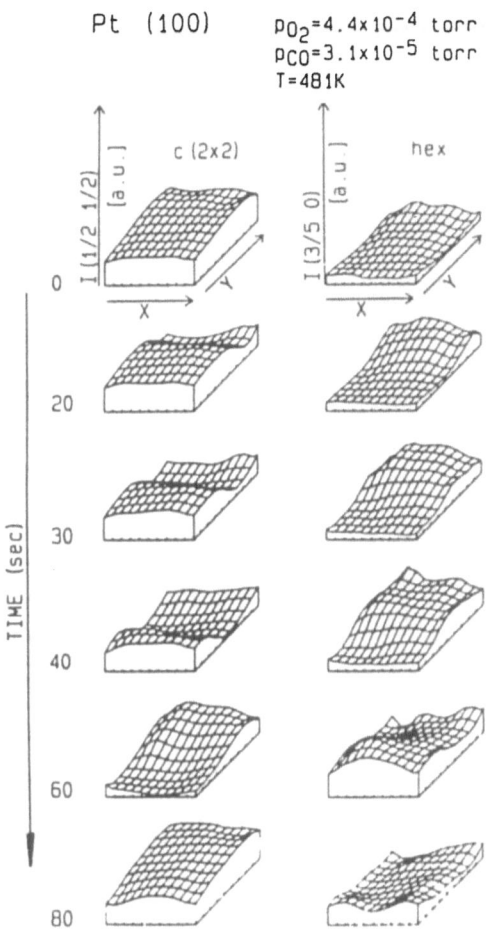

Pt (100) $p_{O_2}=4.4\times10^{-4}$ torr
             $p_{CO}=3.1\times10^{-5}$ torr
             T=481K

Fig. 5.1: Scanning LEED data from a 4x7 mm^2 area of the Pt(100) surface during oscillations, showing the wave-like propagation of the structural transformations of the c(2x2)/ 1x1 and hex phases, respectively. (After [22]).

changes occurring sequentially in Fig. 4.2 are now taking place simultaneously at different parts of the surface. The oscillations on Pt(100) are therefore spatio-temporal and the spatial pattern formation can best be described as a wave-like propagation of the structural changes across the single crystal surface.

Spatial pattern formation on Pt(100) offers a simple explanation for the different characteristics of the LEED and work function measurements shown in Fig. 4.2, where the LEED intensity exhibited very abrupt transitions while the work function varied smoothly. The abrupt changes seen in LEED result from the localized character of these measurements, since the LEED intensity varies sharply as soon as a reaction front passes the region sampled by the LEED beam (diam. ~0.5 mm). The Kelvin probe, on the other hand, sampling over a large region (diam. ~5 mm) shows continuous changes as the sharp changes average out over a large area. This example demonstrates that the effect of spatial pattern formation also appears indirectly in spatially non-resolving methods if they average over differently sized regions.

The spatially resolving Scanning LEED experiments were repeated with a variety of different experimental parameters and in all cases the kinetic oscillations on Pt(100) were associated with spatial pattern formation of macroscopic dimensions [22,24]. A clear connection was found to exist between the spatial organization and the regularity of the oscillations. Regular oscillations occurred when the waves emanated repeatedly from the same edge zone of the crystal, irregular oscillations when the waves changed their direction or when no "triggering center" was identifiable at all. The finding that the waves always emanated from the edge zone of the crystal but never from the center was explained by the higher concentration of structural defects in the edge zone where oxygen can adsorb more easily and which therefore act as a triggering center for the reactive removal of CO.

The observation of concentration waves propagating on the surface appears to be similar in nature to the chemical waves found in non-stirred fluid phase reaction systems. This leads to the idea that like chemical waves, the coupling between diffusion and a reaction step might be the origin of the observed phenomena. The kinetic model of the oscillations on Pt(100) as given by (1) to (4) in Section 4.2.2 had thus to be reformulated as a reaction-diffusion system which was achieved by simply adding a CO diffusion term to (1) which then reads [21]:

$$\frac{\partial u_a}{\partial t} = k_1 a p_{CO} - k_2 u_a + k_3 a u_b - k_4 u_a v_a / a + k_5 \frac{\partial^2 \frac{u_a}{a}}{\partial x^2} \qquad (1)$$

The effect of oxygen diffusion was neglected as this adsorbate is believed to be relatively immobile on the surface compared to CO due to its higher adsorption energy. In order to numerically integrate the system of partial differential equations, a one-dimensional representation of the surface was used, in which 40 compartments were put together in a row and CO was allowed to flow between them following the concentration gradient. The result of the simulation which is shown in Fig. 5.2, shows that the observed spatial pattern formation can be reproduced quite well with the assumption made above.

The mechanism responsible for spatial pattern formation is the coupling between a step with autocatalytic behavior, which is the reactive removal of CO and the diffusion of CO [22,24]. This coupling mechanism gives rise to a propagating reaction front which removes the CO-adlayer. The spatial pattern is primarily created in this step, whereas all subsequent changes on the surface simply follow the primary reaction front with a certain phase shift. The formation of a propagating reaction front can be made plausible in the following way: Consider two adjacent parts of the surface, one where the reactive removal of CO has already started and another where a perfect c(2x2)-CO layer still exists. Following the concentration gradient, some of the CO from the latter part will diffuse to the part which has already started to react and decrease the concentration of CO in the non-reacting part. As the conditions are such that the CO-covered surface is metastable with respect to the oxygen covered surface, a small decrease in the CO-concentration can trigger the reactive removal of CO there too. This process may continue in the same way across the whole surface, thus explaining the formation of a propagating reaction front. But in order to start the whole process, an initial spatial inhomogeneity is needed, where the trigger waves can nucleate. As the waves in the scanning LEED experiments always emanated from the edge zone of the crystal, which is structurally less perfect, we simulated this property in our model calculations by putting an increased number of defects (equivalent to a higher sticking coefficient for oxygen adsorption) into the edge compartment. The simulation depicted in Fig. 5.2 was actually carried out under conditions where the bulk compartments were outside the oscillatory region, but driven from the oscillations in the edge compartment, waves could be triggered in the adjacent bulk compartment which then propagated with constant shape and velocity through the rest of the bulk compartments [21].

The behavior of the oscillating system can become very complex if several triggering centers interact with each other and irregular or even chaotic temporal oscillations may develop. The irregularity of oscillations which dominated in most of the experiments on Pt(100) can be explained by the existence of several active triggering centers from which waves originate independent of each other and then influence each other in such a way that irregular or chaotic oscillations result. Here, it is important to note that due to their non-linear character, chemical waves cannot be superimposed as can electromagnetic waves for example, but they either extinguish each other or only one wave survives. Therefore, even if several

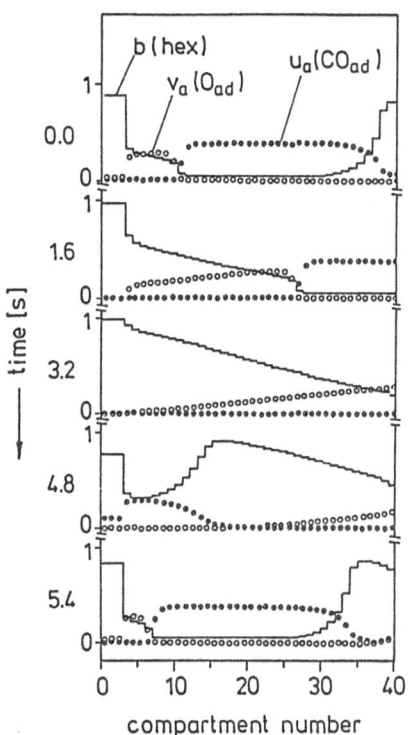

Fig. 5.2: Development of propagating spatial structures in kinetic oscillations on Pt(100) as obtained from numerical integration of eqs. (1)-(4). Forty compartments each of width 10^{-2} cm are used for a one-dimensional representation of the surface. Full line: hex structure (b), full circles: CO coverage (u_a), open circles: oxygen coverage (v_a). (After [22]).

triggering centers exist simultaneously on the surface, one mode may become the dominating one and regular oscillations may again be obtained. This was shown in a two-dimensional simulation of kinetic oscillations on Pt(100) carried out in the so-called cellular automaton technique in which the differential equations of the kinetic model were translated into rules for the various elementary processes on the surface [25]. In these simulations, spatial pattern formation could also be found on spatially uniform surfaces, as density fluctuations in the adsorbate caused by the stochastic nature of adsorption and reaction proved to be sufficient for the nucleation of reaction fronts.

The more realistic case is certainly that of a structurally, non-uniform surface since polycrystalline metals in particular exhibit different surface orientations and many defects. The importance of spatial coupling via CO-diffusion can be demonstrated with the help of the diagram in Fig. 3.4 which shows non-overlapping existence regions for kinetic oscillations on Pt(100) and Pt(110). Under any set of parameters, at most one of the two orientations will oscillate, while the other orientation will remain in a stationary state and either be CO covered or covered by oxygen. Oscillations coupled via CO-diffusion may, however, also evolve in the orientation which, by itself, would be in a stationary state. It is obvious that the macroscopic oscillatory behavior of such a polycrystalline surface is determined to a large extent by the size and distribution of the individual orientations as these properties in turn determine the necessary diffusion lengths for CO.

The same considerations are also important in understanding kinetic oscillations on single crystal surfaces as can be shown with the example of the facetted Pt(110) surface. The LEED results of kinetic oscillations on a strongly facetted surface e.g. the periodic variation of the half-order intensity as shown in Fig. 4.11 very clearly demonstrates that even on this strongly facetted surface, flat (110) regions must exist where the phase transition 1x1↔1x2 takes place. As illustrated by the reaction rate curves in Fig. 4.7, the adsorption properties of the facets and (110) terraces

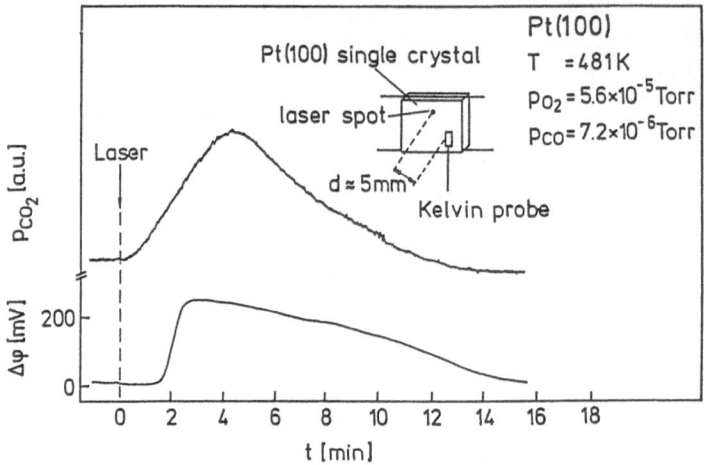

Chemical wave on a Pt(100) surface triggered by a laser pulse. (After [28]).

differ greatly and thus one suspects a coupling via CO-diffusion between the two regions since both regions take part in the oscillations [32]. Admittedly, the interplay between the (110) terraces and the facets, which involves CO-diffusion and mass transport of Pt atoms, is very complex and was not discussed in the simplified mechanism presented in Section 4.3.3.

Using the idea that propagating reaction fronts can emanate from any point of the CO-covered Pt(100) surface maintained in a CO/O_2 atmosphere provided that oxygen gets a chance to adsorb and react there, one can also try to stimulate such an event by locally desorbing CO with a laser pulse. This experiment was conducted with a Pt(100) crystal under reaction conditions corresponding to the borderline of the existence region of oscillations, but where the surface, if left by itself, would have remained CO covered [28]. As shown in Fig. 5.3, the reaction rate starts to increase immediately after the laser pulse which can be attributed to a reaction front that emanates from the spot hit by the laser pulse. This front reactively removes adsorbed CO and behind it an enlarging zone of oxygen covered 1x1 surface is formed which is responsible for the increase in the reaction rate. As shown by the insert in Fig. 5.3, at a distance of ~3 mm away from the spot hit by the laser pulse, a Kelvin probe is used to detect the arrival of the reaction front. In the experiment depicted in Fig. 5.3, the reaction front reaches the Kelvin probe after 1.4 min and causes the work function to rise from the level of the CO covered surface to the level of the oxygen covered surface. It can be shown that the excitation of a single pulse is possible if the p_{O_2}/p_{CO} ratio corresponds to the borderline of the existence region of oscillations. After the pulse has passed, the surface returns to the initial low work function level, whereas oscillations start to develop if the p_{O_2}/p_{CO} conditions correspond to the interior of the existence region for oscillations. Similar experiments, in which propagating reaction fronts were ignited with a laser pulse on Pt films, were carried out by Dath et al. [53] under high pressure conditions. The velocities they obtained for the propagating reaction fronts were of a similar magnitude to the velocities measured with the scanning LEED technique on Pt(100).

The quantitative aspect of the observed phenomena has so far been neglected. Using basic assumptions, Schmidt et al. [42] concluded that CO diffusion under oscillating conditions can hardly account for a spatial coupling over distances of more than 10^{-4} mm. But in such a calculation the essential point has been missed, namely that the coupling between reaction and diffusion can provide a much faster propagating concentration change than with CO diffusion alone [54]. In the numerical simulation of spatial pattern formation on Pt(100) this point was confirmed and by using experimentally determined values for the individual constants it was shown that the experimentally observed velocities of the reaction fronts could in fact be reproduced [21]. The basic assumption in the description of spatial self-organization on Pt(100) therefore appears to be valid. The detailed dependence of the front velocity on the experimental parameters has not been studied due to the irregularities associated with kinetic oscillations on Pt(100). This will be the aim of future research using the technique of stimulating waves with a laser.

5.3 Coupling via Pressure Changes in the Gas Phase

Scanning LEED experiments, which were carried out during kinetic oscillations on Pt(110), could not detect spatially propagating reaction fronts as was the case for Pt(100) [35]. Although deviations from the ideal case of a spatially homogeneously reacting surface exist, the changes appear to take place almost simultaneously in various parts of the surface. Thus a synchronization mechanism different from the one found on Pt(100) must exist on this surface and the basis of this new mechanism is apparent in the measurements of kinetic oscillations on Pt(110) shown in Fig. 3.3. One notices that the periodic variations in the reaction rate are accompanied by small oscillations in p_{CO} which are opposite in phase to the reaction rate and which are of the order of ~1% of p_{CO}. The relation between the reaction rate and the change in p_{CO} follows from the mass balance of the reaction and corresponding changes must also exist for p_{O_2} where, however, they are too small to be detected. The changes in the gas phase, as small as they are, can exert a very strong influence on the surface reaction. This becomes immediately apparent if one considers the very narrow existence region for oscillations on Pt(110) shown in Fig. 3.4.

The idea behind a coupling via the gas phase is that a positive feedback exists between the transition from the CO covered to the oxygen covered surface and vice versa and the concomitant changes in the gas phase concentrations. These surface changes, corresponding to transitions between the low and the high reaction rate branch of the LH-mechanism, occur in the instability region of the reaction (see Fig. 3.1) where the system is sensitive to small variations in the gas phase concentration. If, for example, some part of a CO covered surface has started to react with oxygen, then the concomitant decrease in p_{CO} will enhance the tendency of the surface to move towards the oxygen covered surface and eventually trigger the reactive removal of CO in a surface part which would have remained CO covered without the decrease in p_{CO}. Thus a synchronization mechanism exists which, in contrast to coupling via CO diffusion, affects all parts of the surface in the same way and practically without any delay (although the response of the surface may be associated with some delay).

Since the mechanism which creates spatially propagating reaction fronts in the catalytic CO oxidation works independently of the nature of the substrate, propagating reaction fronts should, in principle, also be found on Pt(110). The picture of spatial self-organization on Pt(110) which is consistent with this requirement and also with a seemingly homogeneously reacting surface, is such that not one but many different reaction fronts are created and triggered simultaneously by changes in the gas phase concentration. As these reaction fronts originate at many different places, either due to defects or coverage fluctuations, they cannot be detected by a technique with a low spatial resolution such as scanning LEED and thus give the impression of a homogeneously reacting surface.

The reasons for the different type of spatial self-organization on Pt(100) and Pt(110), the propagation of reaction fronts in the first case and a spatially homogeneously reacting surface in the latter, can be found in the different widths of the existence regions for oscillations as shown in Fig. 3.4 [35]. As the existence region for oscillations on Pt(100) is relatively broad, concentration changes in the gas phase exert little influence on the behavior of the surface and consequently synchronization can only occur via diffusion of an adsorbate. Surface diffusion will, of course, also occur on Pt(110). Since, however, this surface reacts so much more sensitively to pressure variations, synchronization via gas phase coupling will become the dominant mode here. The different sensitivity of the two surfaces to partial pressure variations can also be shown directly if the crystal surface under oscillating conditions is subject to external periodic perturbations of p_{O_2} or p_{CO}. These measurements, the results of which are discussed in the following Section, showed that p_{O_2} variations with amplitude of more than 4% were necessary before a pronounced response of the Pt(100) surface could be detected, in direct contrast to Pt(110) which already responded to p_{O_2} variations of only 0.2% amplitude [26,30,33]. Because a coupling mode via CO diffusion is inherently more sensitive to perturbations and more easily allows for non-synchronized oscillations in some parts of the surface than coupling via the gas phase, it is easy to see why oscillations on Pt(100) tend to be irregular, while very regular oscillations can be observed on Pt(110).

Spatial pattern formation has also been the subject of several high pressure oscillation studies where the results have been discussed in terms of coupling modes via heat transfer or concentration changes in the gas phase. Spatial inhomogeneities have been detected in the catalytic CO oxidation on a supported Pt catalyst by means of FTIR spectroscopy [3,4,7,8]. Regions of different temperatures on a Pt foil in the catalytic oxidation of H_2 could be imaged directly with infrared thermography [55]. The change in oscillatory behavior when the thermal coupling is modified has been demonstrated in the catalytic CO oxidation on a Pt wire which was cut into two halves thus changing the shape and the existence region of kinetic oscillations [56]. In these studies the coupling between different regions was predominantly mediated through transfer of the reaction heat, but large oscillations in the rate have also been observed with a Pt ribbon under isothermal conditions [57]. Here, the spatial coupling has to be attributed to partial pressure variations in the gas phase. In a cellular automaton study which should simulate the kinetic oscillations found in the catalytic CO oxidation on small Pd particles embedded in a zeolith matrix [58], coupling between different Pd particles was assumed to occur via heat transfer and via CO exchange through the gas phase. Schuster and Gerhardt [59] observed four different modes of spatial pattern formation which were linked to different types of temporal oscillations in the reaction rate. This last example demonstrates very clearly how the macroscopic oscillatory behavior can be determined by spatial self-organization on a catalyst surface.

5.4 Forced Oscillations

Periodic forcing of a system exhibiting autonomous oscillations has been the subject of numerous theoretical and experimental studies [60-62]. The system is subjected to periodic perturbations with the frequency ν_p and the amplitude A and the response of the system characterized by the frequency ν_r is such that it is either phase-locked with a fixed phase difference between perturbation and response or it may be quasiperiodic, in which case the phase difference between perturbation and response changes continuously. As a third possibility chaotic behavior may also be observed. If the system response is phase-locked in a so-called entrainment region, ν_r/ν_p can be expressed as the ratio of two integer numbers such that $\nu_r/\nu_p=\ell/k$ and depending on this value, one distinguishes between subharmonic entrainment for $\ell/k<1$ and superharmonic entrainment for $\ell/k>1$. For $\ell/k=1$, one obtains harmonic entrainment in which the system oscillates with the same frequency as the perturbation.

Several studies which have been carried out with kinetic oscillations in heterogeneously catalyzed reactions suffered from the disadvantage that even the unperturbed system exhibited irregular oscillations [26,58,63-65]. Therefore the response of the system did not show much structure besides a broad harmonic entrainment region. But a very marked behavior of the response is found with kinetic oscillations on Pt(110) which can be ascribed to the excellent regularity and reproducibility of this system. The experimental procedure was such that first kinetic oscillations were adjusted in the high temperature regime and were then subjected to periodic modulations of p_{O_2} with an amplitude of around 1% and a frequency of up to 0.5 s^{-1} [30,33,37].

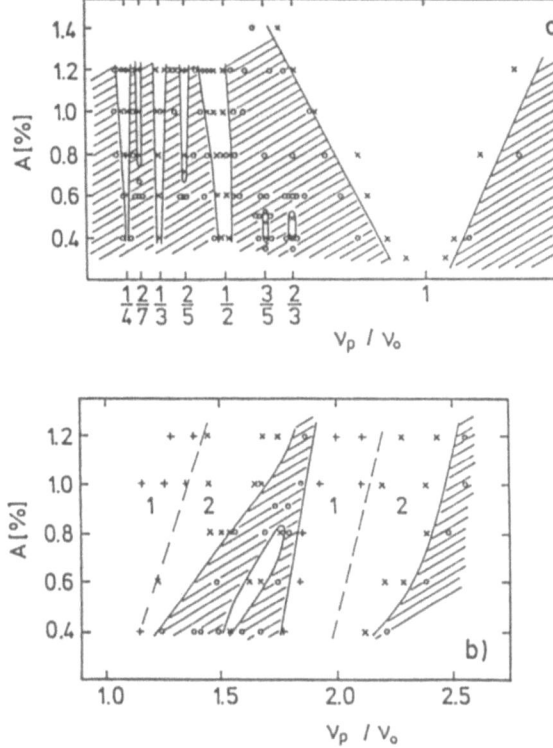

The response of the system can conveniently be summarized in a diagram which, in analogy to a thermodynamic phase diagram, is also termed a phase diagram as it separates a region of well-ordered entrained behavior from regions of a more irregular character where the system responds in a quasiperiodic manner. The different regions were experimentally determined for Pt(110) and are shown in the phase diagram in Fig. 5.4. The parameters, which determine the dynamic response of the system and represent the axis in Fig. 5.4, are the amplitude of the perturbation and the frequency of the perturbation with respect to the frequency of the unperturbed system ν_p/ν_o. The phase diagram exhibits a number of entrainment regions characterized by different ratios ℓ/k. These regions are separated by regions of quasiperiodic behavior which, in Fig. 5.4, are indicated by the shaded areas. A formal analogy has been drawn between such a dynamic phase diagram and the static phase diagram for commensurate and incommensurate structures on a surface, because the ratio of the lattice vectors of the substrate and the overlayer can also be expressed by two integer numbers for commensurate structures and by a rational number for incommensurate structures [37]. We can distinguish between several main regions in Fig. 5.4:

Harmonic entrainment: If ν_p is near ν_o the system oscillates with the same frequency as the perturbation and with a fixed phase difference (phase-locking) with respect to the perturbation. The system behaves similarly to a damped linear oscillator as the amplitude passes through a maximum for $\nu_p=\nu_o$ and the phase shift changes continuously from 0 to π as ν_p is varied. Since the oscillating system is undamped, a finite amplitude is also found outside the resonance region. As expected, the width of the harmonic entrainment region becomes broader with increasing amplitude A.

Sub- and superharmonic entrainement: In total 2, subharmonic (2/3, 1/2) and 7 superharmonic (4/1, 7/2, 3/1, 5/2, 2/1, 5/3, 3/2) entrainment bands were discovered. An example of a complicated temporal pattern is displayed in Fig. 5.5 where

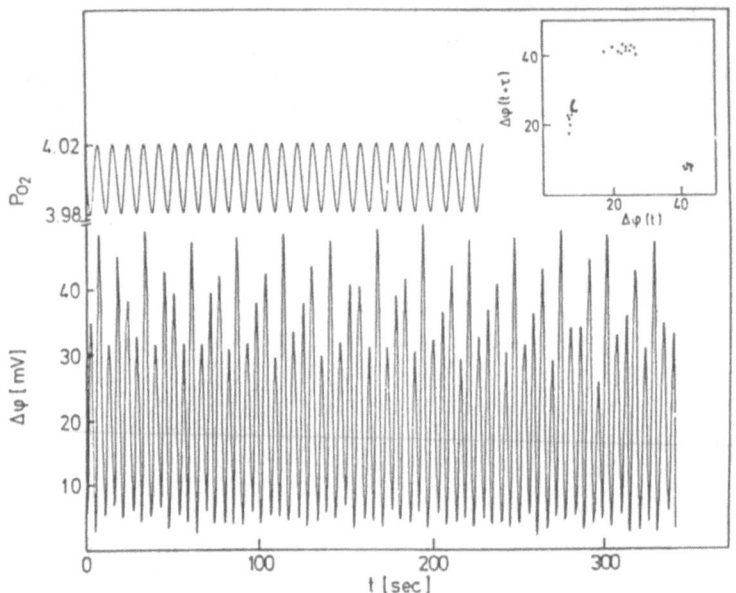

Fig. 5.5.: Time series and stroboscopic plot of 5/3 superharmonic entrainment. (After [33]).

superharmonic entrainment for $\ell/k=5/3$ is shown. It is difficult to distinguish such complicated temporal patterns from quasiperiodic oscillations. This can be solved by means of stroboscopic plots.

An analogy exists with respect to the behavior of a phase transition near the critical temperature T_C which in the "dynamic phase diagram" here corresponds to the borderline of the entrainment region. Near T_C, "critical slowing down" can be observed in a phase transition due to the occurrence of large scale fluctuations or where loosely speaking, it is difficult for the system to decide in which direction to go. "Critical slowing down" in forced oscillations is encountered if one switches on the periodic modulation near the entrainment edge and waits for the system to accommodate for a steady state response. It will take many cycles (up to about 50!) in this region until this situation is achieved compared to only a few cycles needed if the same experiment is carried out well inside the entrainment region.

Forced oscillations on Pt(110) may be regarded as a means of characterizing the dynamic behavior of the system and later on of simulating it with a suitable model, but they are also of direct relevance to the understanding of the oscillatory behavior itself. If one considers oscillations on a heterogeneous surface, synchronized via gas phase coupling, then the situation for each of the different regions subjected to periodic variations of the partial pressures is the same as in an experiment with periodic forcing. Each of the different regions on the surface is likewise to have its own frequency of autonomous oscillations. Its response should be predicted by a dynamic phase diagram of the type shown in Fig. 5.4. If, for example, a region on a Pt(110) surface exists whose own frequency is between 10% and 30% lower than the frequency ν_0 of the rest of the surface, then this region should respond with $1/2 \cdot \nu_0$ to the partial pressure modulation with ν_0 which is exerted by the rest of the surface, as this region is inside the subharmonic entrainment region with $\ell/k=1/2$. Macroscopically, one would expect to observe period-doubling and one might even predict that a transition from simple harmonic oscillations to period doubling may be found as the amplitude of the oscillations becomes smaller. One could then speculate that the transition from ordered behavior to chaos via a series of period doublings observed in the high temperature oscillations on Pt(110), is due to the existence of surface heterogeneities. As the control parameter p_{CO} is varied towards the reaction rate maximum, the amplitude of the oscillations (and hence the partial pressure variations of p_{CO}) becomes smaller, the different regions are no longer synchronized strongly enough but develop their own characteristic response. Significant surface heterogeneities have, in fact, been detected in experiments in which the oscillations at two different regions of the surface were recorded simultaneously by means of two Kelvin probes [69], but at present it is by no means clear whether the mechanism depicted above is actually responsible for the transition to chaos or whether this transition is due to the inherent dynamics of oscillations on a homogeneous Pt(110) surface.

6. Conclusions

The results obtained in a number of single crystal studies show that the analytic tools of surface science can be used to gain a detailed picture of the microscopic mechanism which is concealed by the observed macroscopic phenomena of kinetic oscillations. The results also demonstrate that the apparent simplicity of the chemistry is only superficial, as the energetic interactions of the adparticles gives rise to a series of complex phenomena such as island formation, phase transitions and spatial pattern formation. On the other hand ,it is precisely these characteristics which contribute the new aspects to the study of kinetic oscillations as they are not present in the homogeneous systems such as the BZ-reaction. The principal difference between kinetic oscillations in the homogeneous phase and in a heterogeneous reaction is certainly that the surface inherently represents a system of many individual oscillators which have to be coupled in order to obtain measurable temporal variations. Therefore, the macroscopic behavior is determined by cooperative effects best illustrated by the fact that the oscillatory mechanism on the Pt surfaces itself is also based on a cooperative phenomenon, namely, a surface phase transition.

If one attempts to extrapolate the results of the single crystal studies to kinetic oscillations under atmospheric pressure one could, in principle, expect the structural changes of the surface play an important role. The influence of surface contaminants and non-isothermal conditions is difficult to estimate, however , so that again the only way of arriving at a sound solution is to carry out in situ investigations using suitable experimental techniques. The main emphasis in future projects will probably be on bridging this gap in the reaction conditions and on oscillations on an atomic scale which should be possible using a technique with high spatial resolution such as scanning tunneling microscopy.

ACKNOWLEDGEMENTS

The author would like to thank D. J. Coulman for correcting the English in this article, S. Wasle for the preparation of the drawings and M. Reimers for typing the manuscript.

REFERENCES

1   a) P. Hugo: Ber. Bunsenges. Phys. Chem. **74** (1970) 121;
    b) H. Beusch, P. Fieguth, E. Wicke: Chem. Ing. Techn. **44** (1972) 445

2   L. F. Razón, R. A. Schmitz: Catal. Rev. Sci. Eng. **28** (1986) 89

3   D. J. Kaul, E. E. Wolf: J. Catal. **91** (1985) 216

4   D. J. Kaul, E. E. Wolf: J. Catal. **93** (1985) 321

5   T. H. Lindstrom, T. T. Tsotsis: Surf. Sci. **150** (1985) 487

6   a)V. A. Burrows, S. Sundaresan, Y. J. Chabal, S. B. Christman: I: Surf. Sci. **160** (1985) 122, II: Surf. Sci. **180** (1987) 110;
    b) N. A. Collins, S. Sundaresan, Y. J. Chabal: Surf. Sci. **180** (1987) 136

7   R. Sant, E. E. Wolf: J. Catal. **110** (1988) 249

8   F. Schüth, Thesis, Univ. Münster (1988)

9   J. E. Turner, B C. Sales, M. B. Maple: Surf. Sci. **103** (1981) 54

10 Experimental details can be found in refs. 23 and 24 . The general background of the experimental techniques used in surface chemistry is described in refs. 11 and 12.

11 G. Ertl, J. Küppers, *Low Energy Electrons and Surface Chemistry*, Verlag Chemie, Berlin 1985

12 See, e.g. R. P. H. Gasser, *An Introduction to Chemisorption and Catalysis by Metals*, Clarendon Press, Oxford 1985

13 H. Niehus, G. Comsa: Surf. Sci. **102** (1981) L14

14 R. C. Yeates, J. E. Turner, A. J. Gellman, G. A. Somorjai: Surf. Sci. **149** (1985) 175

15 T. Engel, G. Ertl: Advan. Catal. **28** (1979) 1

16 a) M. A. Van Hove, R. J. Koestner, P. C. Stair, J. P. Bibérian, L. L. Kesmodel, I. Bartoš, G. A. Somorjai: Surf. Sci. **103** (1981) I 189; II 218;
    b) W. Moritz, Habilitationsschrift, Univ. München (1984);
    c) R. J. Behm, W. Hösler, E. Ritter, G. Binnig: Phys. Rev. Lett. **56** (1986) 228

17 a) H. Niehus: Surf. Sci. **145** (1984) 407;
    b) T. E. Jackman, J. A. Davies, D. P. Jackson, W. N. Unertl, P. R. Norton: Surf. Sci. **120** (1982) 389

18 G. Ertl, P. R. Norton, J. Rüstig: Phys. Rev. Lett. **49** (1982) 177

19 P. R. Norton, P. E. Bindner, K. Griffiths, T. E. Jackman, J. A. Davies, J. Rüstig: J. Chem. Phys. **80** (1984) 3859

20 M. P. Cox, G. Ertl, R. Imbihl, J. Rüstig: Surf. Sci. **134** (1983) L517

21 R. Imbihl, M. P. Cox, G. Ertl, H. Müller, W. Brenig: J. Chem. Phys. **83** (1985) 1578

22 M. P. Cox, G. Ertl, R. Imbihl: Phys. Rev. Lett. **54** (1985) 1725

23 M. Eiswirth, R. J. Schwankner, G. Ertl: Z. Phys. Chem. N.F. **144** (1985) 59

24 R. Imbihl, M. P. Cox, G. Ertl: J. Chem. Phys. **84** (1986) 3519

25 P. Möller, K. Wetzl, M. Eiswirth, G. Ertl: J. Chem. Phys. **85** (1986) 5328

26 R. J. Schwankner, M. Eiswirth, P. Möller, K. Wetzl, G. Ertl: J. Chem. Phys. **87** (1987) 742

27 R. F. S. Andradé, G. Devel, P. Borckmans: to be published

28 T. Fink, R. Imbihl, G. Ertl: submitted to J. Chem. Phys.

29 M. Eiswirth, G. Ertl: Surf. Sci. **177** (1986) 90

30 M. Eiswirth, Thesis, Univ. Munich (1987)

31 S. Ladas, R. Imbihl, G. Ertl: Surf. Sci. **197** (1988) 153

32 S. Ladas, R. Imbihl, G. Ertl: Surf. Sci. **198** (1988) 42

33 M. Eiswirth, G. Ertl: Phys. Rev. Lett. **60** (1988) 1526

34 M. Eiswirth, K. Krischer, G. Ertl: Surf. Sci. **202** (1988) 565

35 M. Eiswirth, P. Möller, K. Wetzl, R. Imbihl, G. Ertl: J. Chem. Phys. **90** (1989) 510

36 M. Eiswirth, P. Möller, G. Ertl: Surf. Sci. **208** (1989) 13

37 M. Eiswirth, G. Ertl: Appl. Phys. **A47** (1988) 91

38 a) R. Imbihl, M. Sander, G. Ertl: Surf. Sci. **204** (1988) L701;
    b) A. L. Vishnevskii, V. I. Savchenko: React. Kinet. Catal. Lett. **38** (1989) 187

39 a) M. Ehsasi, J. H. Block, K.Christmann, W. Hirschwald: J. Vac. Sci. Technol. **A5** (1987) 821;
    b) M. Ehsasi, Thesis, FU Berlin (1989).

40 M. Ehsasi, C. Seidel, H. Ruppender, W. Drachsel, J. H. Block, K. Christmann: Surf. Sci. **210** (1989) L198

41 S. Ladas, R. Imbihl, G. Ertl: Surf. Sci., in press

42 S. B. Schwartz, L. D. Schmidt: Surf. Sci. **183** (1987) L269

43 S. B. Schwartz, L. D. Schmidt: Surf. Sci. **206** (1988) 169

44 R. J. Behm, P. A. Thiel, P. R. Norton, G. Ertl: J. Chem. Phys. **78** (1983) 7437; 7448

45 J.-P. Dath, J.-P. Dauchot: J. Catal. **97** (1986) 100

46 a) G. Pirug, G. Brodén, H. P. Bonzel: Proc. 7th Int. Vacuum Congr. Vienna (1977), p. 907; Chem. Phys. Lett. **73** (1980) 306;
    b) P. R. Norton, K. Griffiths, P. E. Bindner: Surf. Sci. **138** (1984) 125

47 N. Freyer, M. Kiskinova, G. Pirug, H. P. Bonzel: Surf. Sci. **166** (1986) 206

48 M. Flytzani-Stephanopoulos, L. D. Schmidt: Progr. Surf. Sci. **9** (1979) 83

49 a) C. E. Smith, J. P. Biberian, G. A. Somorjai: J. Catal. **57** (1979) 426;
    b) H. Niehus, G. Comsa: Surf. Sci. **93** (1980) L147

50 A. K. Galwey, P. Gray, J. F. Griffiths, S. M. Hasko: Nature **313** (1985) 668

51 a) G. Padberg, E. Wicke: Chem. Eng. Sci. **22** (1967) 1035;
    b) N. I. Jaeger, K. Möller, P. J. Plath: J. Chem. Soc. Faraday **82** (1986) 3315;
    c) J. R. Brown, G. A. D'Netto, R. A. Schmitz: in *Temporal Order* (Eds. L. Rensing and N. I. Jaeger), Springer, Heidelberg 1985, p. 86;
    d) H. U. Onken, Thesis, Univ. Münster (1987);
    e) H. U. Onken, E. Wicke: Ber. Bunsenges. **93** (1989) 191

52 R. J. Behm, P. A. Thiel, P. R. Norton, P. E. Bindner: Surf. Sci. **147** (1984) 143

53 J.-P. Dath, J.-P. Dauchot: J. Catal., J. Catal. **115** (1989) 593

54 a) R. J. Field, H. Burger (Eds.), *Oscillations and Travelling Waves in Chemical Systems*, John Wiley, New York 1985; b) J. Tilden: J. Chem. Phys. **60** (1974) 3349

55 G. A. D'Netto, J. R. Brown, R. A. Schmitz: IChemE Symp. Ser. **87** (1984) 247

56 P. K. Tsai, M. B. Maple, R. K. Herz: J. Catal. **113** (1988) 453

57 L. F. Razón, S.-M. Chang, R. A. Schmitz, Chem. Eng. Sci. **41** (1986) 1561

58 N. I. Jaeger, P. J. Plath, P. Svensson, in *Spatial Inhomogeneities and Transient Behavior in Chemical Kinetics*, (Ed. G. Nicolis), in press

59 a) M. Gerhardt, Thesis, Univ. Bremen (1987); b) H. Schuster, Thesis, Univ. Bremen (1987); c) A. K. Dewdney: Scientific American **259** (1988) 86

60 See, e.g. a) H. Haken; *Advanced Synergetics*, Springer Series in Synergetics, Vol. 20, Springer-Verlag, Berlin, Heidelberg 1983; b) J. M. T. Thompson, H. B. Stewart; *Non-linear Dynamics and Chaos*, Wiley, New York, 1986

61 T. Kai, K. Tomita: Progr. Theor. Phys. **61** (1979) 54

62 F. W. Schneider: Ann. Rev. Chem. **36** (1985) 347

63 M. Marek, in *Temporal Order* (Eds. L. Rensing and N. I. Jaeger) Springer, Berlin (1985), p. 105

64 H. U. Onken, E. Wicke: Ber. Bunsenges. Phys. Chem. **90** (1986) 976

65 S. C. Capsaskis, C. N. Kenney: J. Phys. Chem. **90** (1986) 4631

66 D. T. Lynch, G. Emig, S. Wanke: J. Catal. **97** (1986) 456

67 W. Hösler, E. Ritter, R. J. Behm: Ber. Bunsenges. Phys. Chem. **90** (1986) 205

68 J. Falta, R. Imbihl, M. Henzler: submitted to Phys. Rev. Lett.

69 R. Imbihl, S. Ladas, G. Ertl: Surf. Sci. 215 (1989) L307

Optimization of Heterogeneous-Catalyst Structure: Simulations and Experiments with Fractal and Non-Fractal Systems

D. Avnir, O. Citri, D. Farin, M. Ottolenghi, J. Samuel, and A. Seri-Levy

Institute of Chemistry and the F. Haber Research Center for Molecular Dynamics, The Hebrew University of Jerusalem, Jerusalem 91904, Israel

We explore the effects that the geometric details of heterogeneous catalysts have on their performance, in order to suggest guide-lines for the design of catalysts with optimal structures. The following geometric parameters are investigated: *(a)* The (fractal and non-fractal) distribution pattern of surface active sites; *(b)* the particle size of the catalyst; *(c)* the surface fractal dimension; *(d)* the average pore size.

1. Background

Heterogeneous catalysis is perhaps the most prominent example in chemistry of the important role that geometry plays in determining reaction efficiencies [1]. Examples of well-defined geometries include crystal planes [2] and the pore network of zeolites [3]. Yet the majority of industrial and laboratory-made catalysts are structurally very complex: the surfaces are irregular, the pore-size has a typical distribution as does the particle size, and the surface is not usually homogeneously reactive at all its sites.

The recognition that a wide variety of randomly irregular structures in Nature is describable in terms of fractal geometry [4,5] has boosted intensive interest in the chemical reactivity of fractal systems [6-10] including catalytic processes [12-21]. Here, we briefly describe some of our contributions to the latter topic, with specific emphasis on the following structural parameters: the active sites distribution; the particle size; the surface fractal dimension; and the pore size. Preliminary guide-lines for the optimization of these structural parameters are presented. Detailed accounts of the material sketched in this brief report, follow in subsequent publications. A detailed account of size effects (Section 3) appeared in [10,19-21].

2. The Optimization of the Activity Zones Distribution

Intensive activity in the development of nanostructure technology [22] and reports on the ability to design surface features even on a scale of tens of angstroms [23] now permit the design of catalyst structures, from the macroscopic down to the molecular scale, both of the geometric features on the surface and of the distribution of active zones on it. We deal with the former in Section 4, and here we discuss the latter.

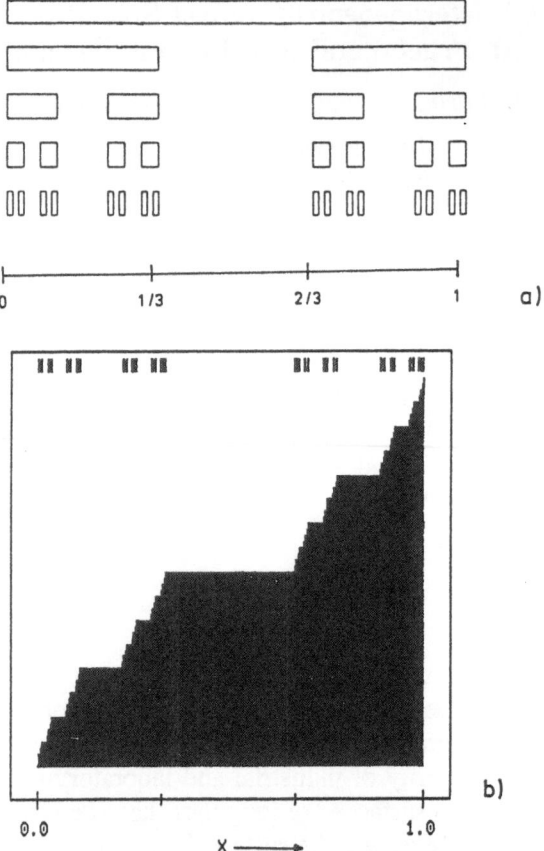

Fig 1: *(a)* The construction of a Cantor set and *(b)* 'Devil staircase' derived from it [4a]

There are two aspects to the question of optimizing the distribution pattern of active zones. The first relates to the question of the distribution of supported catalysts on an inert support: there are a number of indications that the growth of catalytic zones results in fractal patterns [12,13,24] and not in a homogeneous distribution. The second is the pattern of imperfections on a single metal crystallite. Here, there is growing evidence, both theoretical [25] and experimental [26,27] that a common form of imperfection on the surface of such crystals is that of a fractal Cantor set, or Devil Staircase, both shown in Fig. 1.

With this background in mind, let us ask the following question: assuming that the activity is confined to a collection of catalytic sites which form a certain shape, e.g., the Cantor set; how does this specific distribution affect the kinetics of catalytic reactions with it? Is this 'natural' pattern of imperfections the optimal form?

We answer these questions for the simple catalytic process

$$A \rightarrow P \tag{1}$$

in which A diffuses from the bulk solution to the surface and reacts with the catalytic sites. In Section 4, we use random walk simulations in order to answer these questions on a molecular level; here we treat this question on a macroscopic scale (e.g., a patterned electrode [28]). We do so by solving the classical diffusion/reaction equations, which, for Reaction 1 are:

$$\frac{\partial A}{\partial t} = -k(x)A + K_A \nabla^2 A \tag{2}$$

$$\frac{\partial P}{\partial t} = +k(x)A + K_P \nabla^2 P \tag{3}$$

where A and P are concentrations, k is the reaction constant (which vanishes to zero at a small distance from the interface) and K_A, K_P are the diffusion constants. The simulations are carried out for a one-dimensional catalytic wall in a two-dimensional array of $n \times m$ equal cells. The reaction takes place in certain cells of the first row (i.e., those cells which are near the wall). In these active cells, both terms on the right hand side of (2) and (3) are calculated, while in the rest of the cells only diffusion takes place. The Laplacian is calculated by fast Fourier transform and the time propagation by the Euler method (for more technical details, see [29,30]).

Before considering questions of optimal distribution of sites, we draw attention to an important feature of reactions with heterogeneous surfaces: A very complex pattern of concentration profile, of both reactants and products, is developed near the surface (shown in Fig. 2 for a Cantor set). It is clearly evident that the kinetic

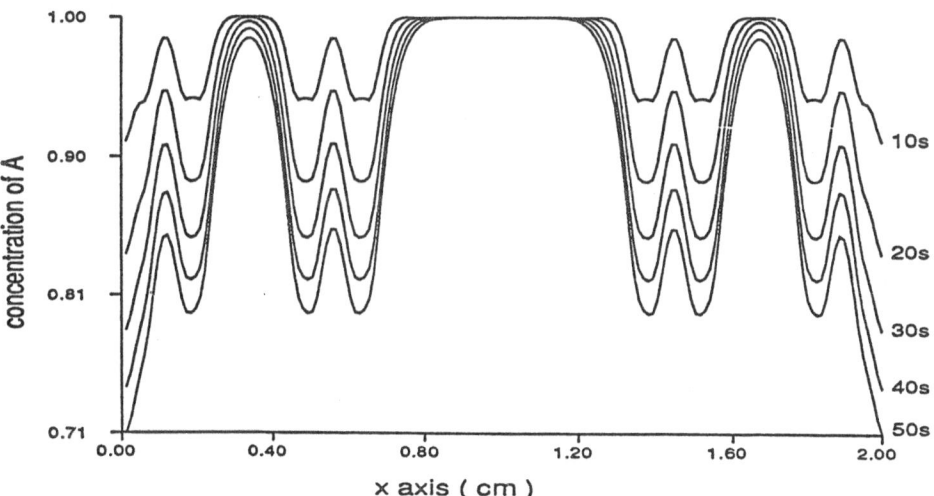

Fig 2: Evolution of concentration profiles, generated by a reaction on a Cantor interface, at the third row parallel to interface (0.03 mm from interface) at intervals of 10 sec (numbers on right side indicate the time). $k=0.1$ sec^{-1}, $K_A=5\times10^5$ cm^2/sec, $K_P=1\times10^5$ cm^2/sec. The Cantor set is generated by four iterations. The grid size 162×20 pixels.

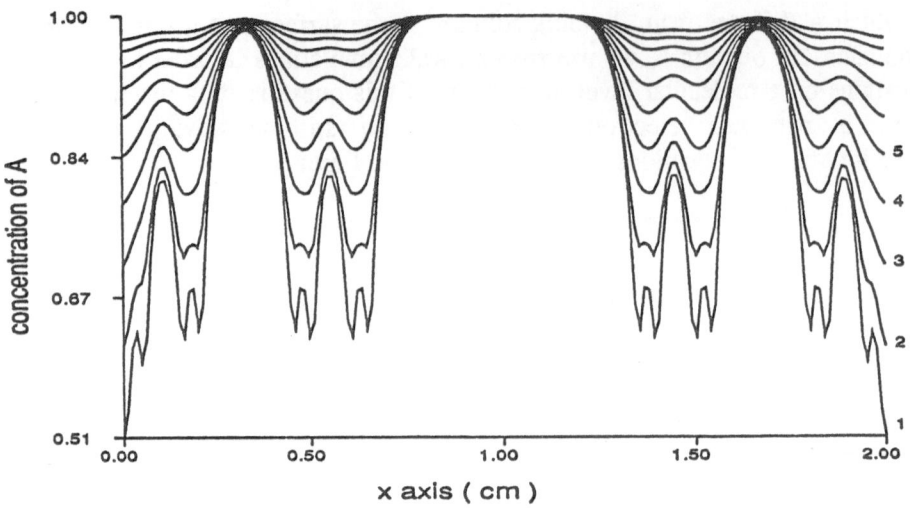

Fig 3: Evolution of concentration profiles at ten first rows (0.01 mm apart) parallel to interface at t=50 sec. Numbers on the right side indicate the row number. Parameters as in Fig. 2.

analyses of catalytic processes on heterogeneous surfaces, cannot be performed as is the common practice, by treating the problem as a homogeneous one. We notice that a reaction which takes place on such a patterned surface, carries away with it, through diffusion, structural information in the form of complex three-dimensional concentration profiles of reactants and products. In other words, surface memory effects are operative in such systems, at a certain distance (Fig. 2) or a certain time (Fig. 3), the surface pattern is built in the bulk and then dissipated away [31,32]. A detailed account of those memory effects will be presented elsewhere.

We studied three types of distributions: a uniform distribution (the active zones are equally spaced), a fractal Cantor set gap distribution (the active zones form a Cantor set) and a configuration where there is one active zone at the center of the interface. The total active area (number of active cells) in all the distributions is kept the same (in the simulations presented here, there are 32 active cells out of 162 'wall' cells). Figures 4a and 4b show the accumulation of P as a function of time (i.e., the conversion efficiency) for the three different distributions, for two k values. As can be seen, the most efficient distribution is the uniform one while the less efficient is the concentrated one.

The reason for this behaviour is the following interplay between reaction and diffusion: as the reaction starts and the solution near the interface is depleted of A, diffusion from inactive zones supplies 'fresh' A molecules to the active zones. While in the uniform distribution, each active zone is surrounded by inactive zones which can supply fresh A, in the concentrated configuration there is a big zone of depleted A and new molecules can diffuse into this area from two sides only. The fractal configuration, which consists of pairs of active zones, falls between these

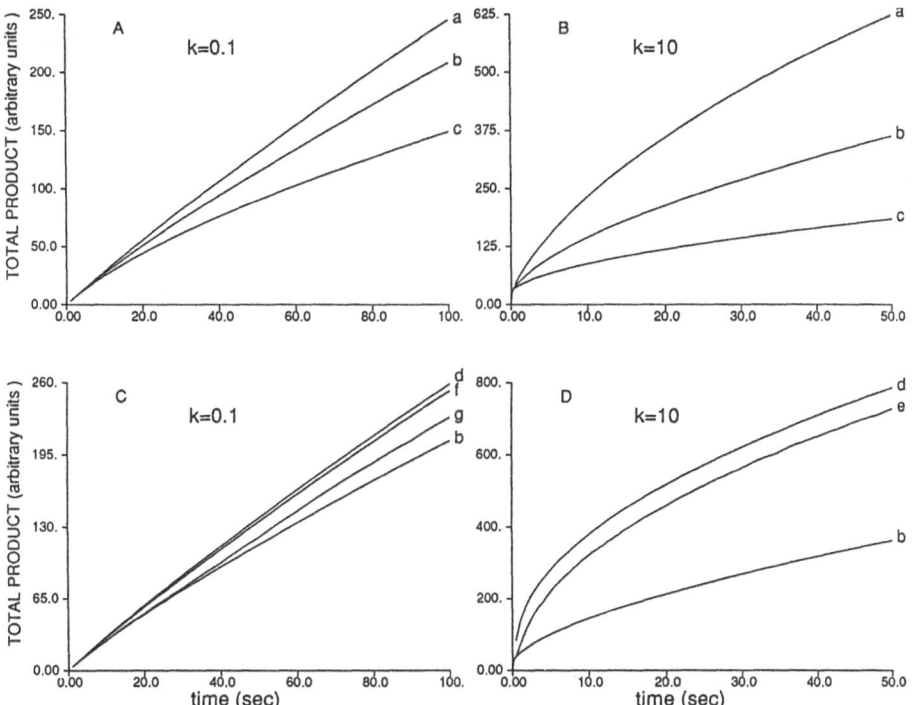

Fig. 4: The effects of active-zones distribution on reaction efficiency. The total accumulation of product is shown as a function of time. The active pattern occupies in all cases 32 out of 162 pixels. Diffusion constant: $5 \cdot 10^{-5}$ cm^2/sec. A ($k=0.1$ sec^{-1}) and B ($k=10$ sec^{-1}) show the results for the following fixed patterns: (a) uniform distribution; (b) fractal Cantor set distribution (c) all active pixels are centered. C ($k=0.1$ sec^{-1}) and D ($k=10$ sec^{-1}) show the effects of randomization of the Cantor pattern at the following rates (randomization/sec): (d) 10; (e) 2; (f) 1; (g) 0.1; (b) 0 (as in A,B).

two. Comparison of Fig. 2 with Fig. 5 clarifies this situation further. Figure 5 shows the concentration profile of A near a uniform patterned wall, at equally spaced distances from the wall, 50 sec after reaction start. As can be seen, the 'modulation' of the concentration profile near the uniform wall (Fig. 5) is less enhanced than near the fractal wall (Fig. 2) and the concentration profiles become homogeneous at quite a small distance from the interface.

The pattern of the collection of reactive sites is, in many instances, dynamic. This is due to either restructuring of the surface metal atoms, or to reversible poisoning which at a (quasi)-steady state, occupies a certain percentage of the active sites. We studied the effects of that dynamic behaviour under the condition $dD/dt=0$, i.e., the specific location of the active sites is randomized, but it is always a Cantor set with the same D. The results are seen in Figs 4c and 4d: Randomization increases the efficiency of the reaction, and the effect is more pronounced the higher

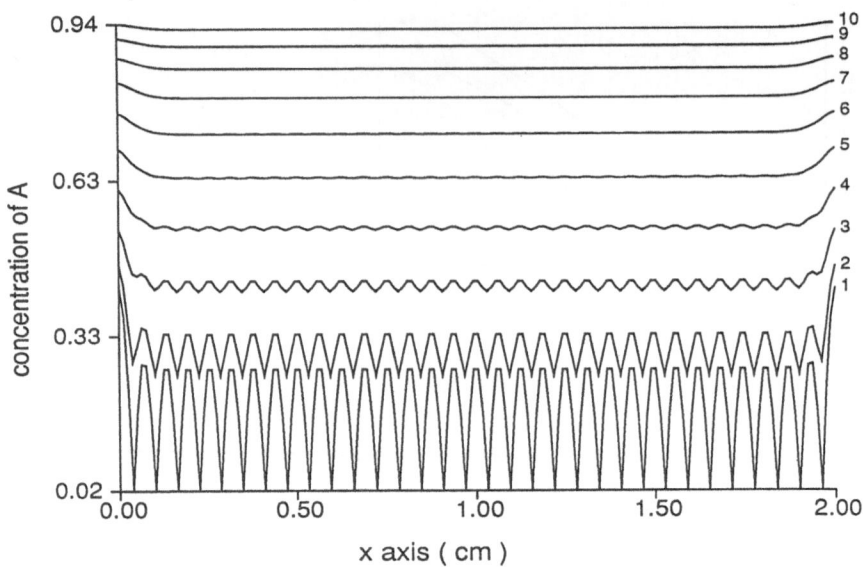

Fig. 5: Concentration profiles of A at ten first rows parallel to the interface (0.01 cm apart) at t=50 sec, near a uniform distribution wall. Numbers on right indicate the row number. k=10 sec^{-1}, K_A=5×10^{-5} cm^2/sec.

the randomization rate. This phenomenon can be understood by realizing that after each randomization step, the active sites are exposed to non-depleted regions of the solution of A molecules.

Finally we notice that sensitivity to patterns of active sites is greater for the faster reaction (Fig. 4). The two limiting factors of sensitivity are, on the one hand, extremely slow reactions (equivalent in their effects to efficient stirring) for which the reaction is insensitive to the surface pattern; and diffusion-limited reactions, on the other hand, for greatest sensitivity.

3. Optimization of the Particle Size

Of the various geometric effects discussed in this report, the one which has been explored in great depth, is that of size, both in our laboratory [19-21,33] and in others [12,18,34]. We briefly highlight some of the main conclusions of that study relevant to catalysis.

First we notice that the fractality of rough surfaces can be elucidated from area/size analysis [34e,34f,35]. Figure 6 illustrates the reason: one can determine the fractal dimension of the boundary line of the largest particle by the usual technique of length vs yardstick-size analysis; an equivalent procedure is to determine the dependency of the boundary length, *l*, as measured with some fixed yardstick (e.g., one pixel, N_2, etc.), on the diameter, 2R, of the particle. The collection of particles in Fig. 6 is such that one iteration is added from the smallest to the largest; hence:

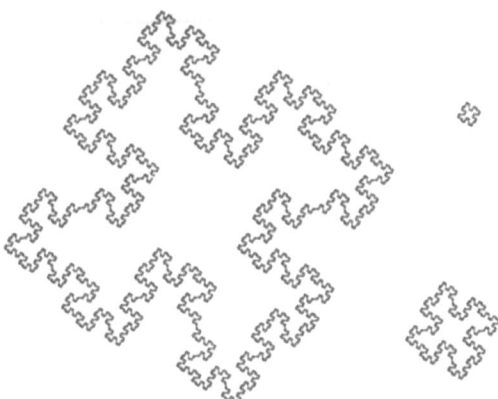

Fig 6: Three particles of the same surface fractal dimension. Notice that one iteration is removed with each decrease in size.

$$l \propto R^D \tag{4}$$

which, for a smooth circle, gives $D = 1$; or for area, S, measurements:

$$S \propto R^D \tag{5}$$

where $D = 2$ refers to a smooth surface sphere. In those catalytic processes where the activity, a, is first-order in the number of active sites, n, (most reactions), $a \propto n \propto S$. Therefore, if all surface sites react, then

$$a \propto R^D \quad \text{(per particle)} \tag{6}$$

For instance, if $D = 2$, then a 'structure-insensitivity' [36] situation is obtained. However, more often than not, only a sub-set of sites reacts. This sub-set may still obey a relation similar to (5), but as it refers now to a reaction analysis, we shall use the symbol D_R, the reaction dimension:

$$a \propto R^{D_R} \tag{7}$$

$$a_g \propto R^{D_R - 3} \quad \text{(per gr [34e,f])} \tag{8}$$

$$a_t \propto R^{D_R - 2} \quad \text{(turn over units per exposed surface site)} \quad , \tag{9}$$

Indeed, it was found that very many catalytic processes obey (7-9). (A detailed account is given in [19-21,33,34]).

The question of optimization of particle size, can be divided into two parts. The first one deals with non-porous objects of relatively smooth surfaces (D=2), for which (8,9) hold. This is the case, for instance, for catalysis on dispersed metals [36]. Equation (7) tells us that if a single particle is considered, the activity will always be larger, the larger the size of the particle. In practice, however, one often uses a certain weight of a catalyst. For optimization of the activity of a unit weight of a catalyst, we see from (8) that $D_R > 3$ and $D_R < 3$ cases should be treated differently: For the $D_R<3$ cases, one should aim for a reduction in catalyst size, whereas for $D_R>3$ situations, one should aim for the opposite. The majority of cases fall into the first category, and only a few cases are known with $D_R>3$. An example of the latter is ammonia synthesis on Fe dispersed on MgO, for which $D_R=5.8$ was observed.

The second question of optimization deals with those cases where the object has a rough surface with $D\geq2$. For these cases it can be shown [10,37] that m in the empirical relation:

$$a \propto S^m \qquad (10)$$

in which S is the N_2-BET surface area and m the reaction order in surface area, is

$$m=(D_R-3)/(D-3) \qquad (11)$$

From (10,11), which are a generalization of Wenzel's Law [10] it is apparent that reaction efficiency can be increased by reducing the particle size for $D_R<D$, $m>1$ cases, and by increasing the particle size for $D_R>D$, $m<1$ reactions. To understand this conclusion, we recall that the dependency of the total surface area, S, on R is composed of two contributions: the outer particle surface area, which depends strongly on R, and the hidden inner area, which is less sensitive to R (at the extreme, S of highly porous materials is independent of R). Consequently, if a reaction takes place preferentially in internal parts, it will show up in $m<1$. It is thus possible to increase the efficiency of such a reaction by increasing the relative contribution of these internal parts, i.e., by increasing the particle size. The other situation, $D_R<D$, $m>1$, indicates the opposite picture, namely, either geometrical screening of the internal parts of the surface by the external ones, or chemical selectivity at the surface. The explanation is that: if screening operates, then the outer surface becomes the important factor, and this outer surface is strongly dependent on the inverse of the size, i.e., strongly dependent on S. Similarly, the relative weight of a population of a sub-set of reactive sites (compared to all surface points) increases with a decrease in R (i.e., increase of S).

Particle size considerations for optimization of catalytic processes, are demonstrated by the following case: Curtis et al have studied the kinetics of liquefaction of coal, asphalthene and preasphalthene on the industrial catalyst Shell 324 $NiMo/Al_2O_3$, in order to detect the role of diffusion in the pores of the catalyst on its performance [38]. This was done by measuring the effect that the particle size of the catalyst has on the rate-determining constants for each of the processes. It was very interesting to find that the rate constants obey (12) with very low D_R values, in the range 2.0-2.2. (Fig. 7. See caption for some details). These low D_R values are an indication that only the *outer* regions of the catalyst contribute significantly to the kinetics, i.e., here one has the simple radius/area relation of a sphere. This conclusion is in agreement with Curtis et al's own conclusion that there exist intra-pore diffusional limitations. Thus, our practical advice would be to abandon efforts to increase the surface area through porosity, in favor of producing non-porous particles of a very small size.

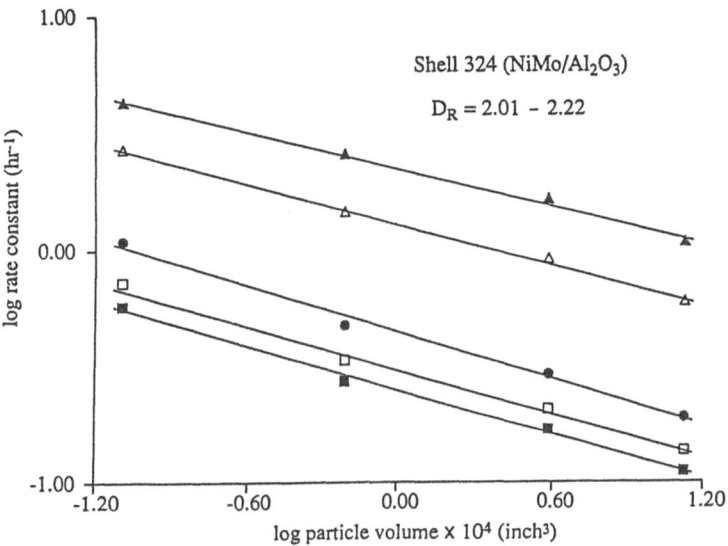

Fig 7: The dependence of liquefacation rate constants on the catalyst particle volume ($a \propto$ volume$^{(D_R-3)/3}$; c.f. eq. (8)). C=Coal, P=Preasphalthene, A=Asphalthene, O=Oil. ■, ▲: k_1, k_2, respectively of the C process $C \xrightarrow{k_1} P \xrightarrow{k_2} A$. □, △: k_{-1}, k_2, respectively of the P process $C \xleftarrow{k_{-1}} P \xrightarrow{k_2} A$. ●, k_3 in the A process $A \xrightarrow{k_3} O$.

4. Optimization of the Surface Fractal Dimension [41]

In Section 2, we studied reaction (1) on a macroscopic level, and a fractal set distributed on a straight line. In this Section we demonstrate the effects of surface irregularity on a molecular scale. Consequently, whereas in Section 2 we used the method of numerical solution of the diffusion/reaction equations, here, we use random-walk simulations according to the Monte-Carlo method.

In order to explore the effects of surface irregularity, we use the set of fractal lines shown in Figs 6,8,9. An initial reservoir of 10^4–10^5 A molecules was distributed homogeneously within an area limited by the fractal line on one side, and by a line at a distance of 89 pixels from it on the other side (Fig. 9). Reaction (1) was then carried out under diffusion-controlled conditions (reaction upon collision), and $P(t)/A_0$ (the conversion) monitored. An important result is shown in Fig. 10: If the ratio between the end-to-end length of the fractal object (shown in Fig. 9) and the reacting-area above the line is kept constant, (1/160 in our simulations) a marked effect of the surface irregularity is observed: The higher the D value, the more efficient is the catalytic reaction. One can understand this observation by noticing that under the condition of the above fixed ratio, more of the reactive line is avail-

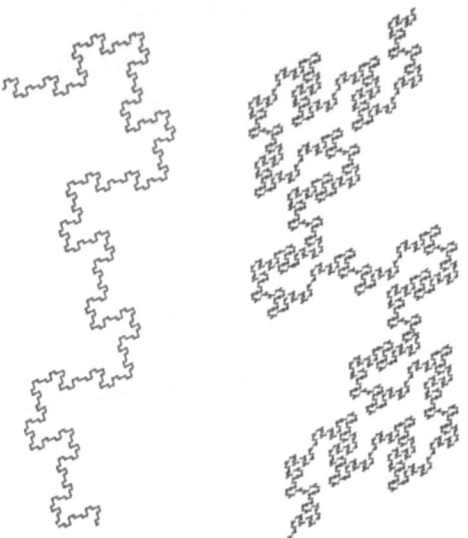

Fig. 8: The fractal lines used for the simulations: a) D=1.61 (top); b) D=1.50 (Fig 6); c) D=1.34 (bottom); d) D=1.33 (Fig. 9); e) D=1.00 (a straight line).

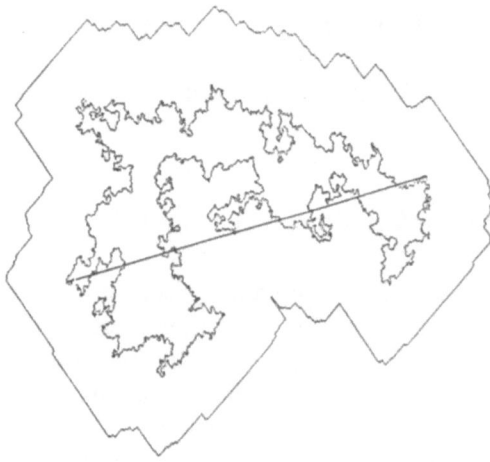

Fig. 9: A random fractal encapsulated within the reactants reservoir. The size of the object is its largest linear extent.

able for the reacting molecules, the higher the D value. Experimentally, this is equivalent to changing D for a fixed particle radius, R; We have seen in the previous Section in equation eq. (7) that under such conditions, activity increases with D.

It has been predicted [17,18,39,40] that the kinetics of Reaction (1) under the simulation conditions used here, is describable in term of the equation

$$P(t)=k^* t^{(2-D_R)/2}$$

(12)

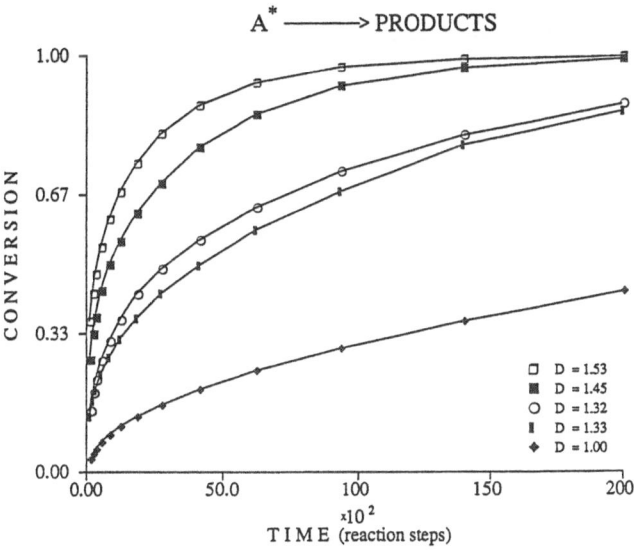

Fig. 10: The conversion as a function of time for the various surface irregularities shown in Figs 6,8,9, for a fixed object-size/reservoir area = 1/160.

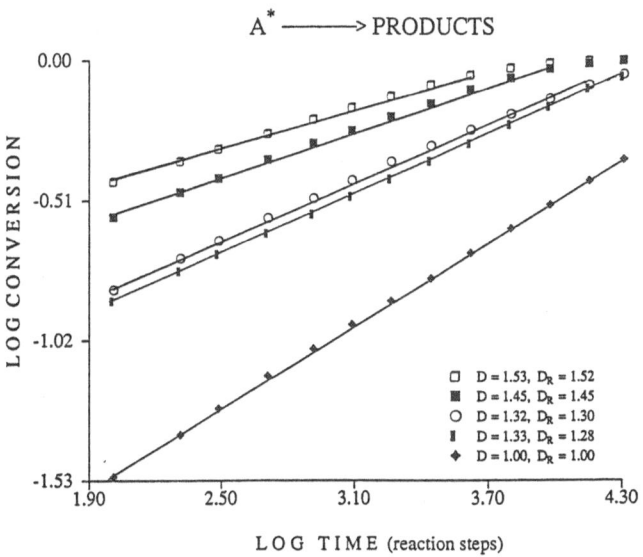

Fig. 11: The same as Fig. 10, analysed according to eq. (12).

where k^* is a constant (discussed elsewhere [21b,31,41]). Indeed, the data in Fig. 10 obeys this equation nicely, as is shown in Fig. 11: The agreement between the simulation D_R values and the D values of the lines, is good.

Interestingly, when k^* in (12) is taken as fixed for the various fractal lines used here, an *opposite* dependency on D is observed (Fig. 12): Here, the smoother

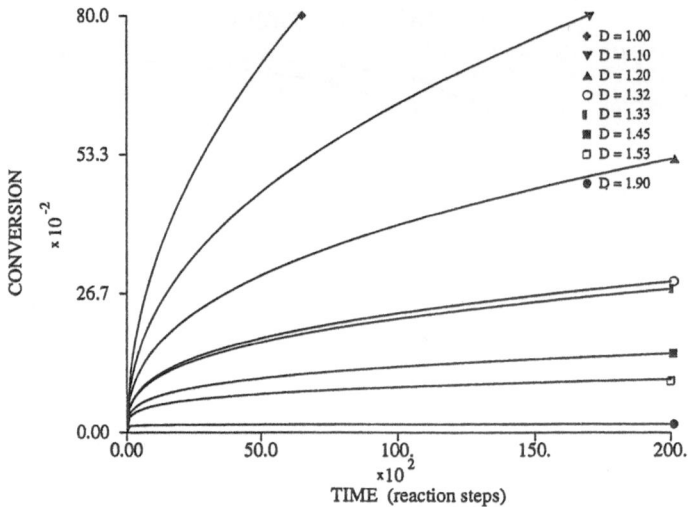

Fig. 12: The same as Fig. 10, but under the condition of fixed k^* (eq. (12)) for all fractal lines. Notice the opposite behaviour compared to Fig. 10.

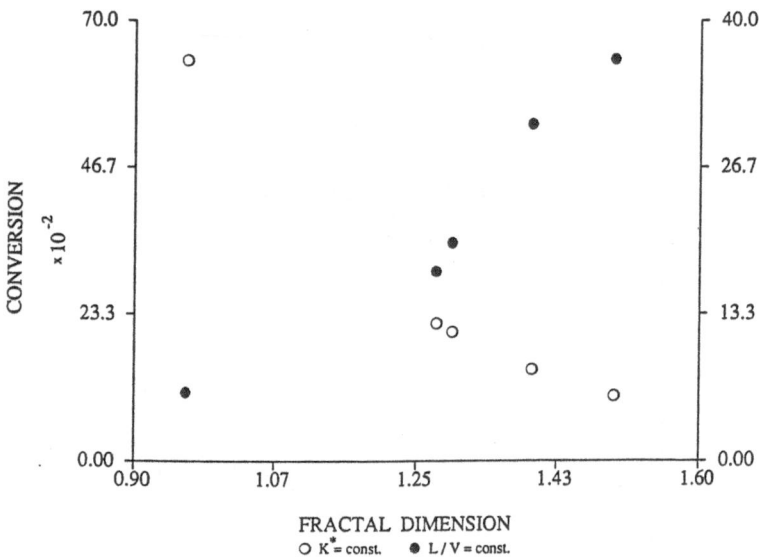

Fig. 13: The effect of surface fractal dimension on reaction efficiency under the conditions of Figs 10 and 12. For comparison, the conversion after 1000 reaction steps was taken.

the surface, the more efficient the reaction. Elsewhere it is shown [31,41] that fixing k^* amounts to more or less fixing the total length of the fractal line, which is experimentally equivalent to fixing the surface area. Here, the reason for the effect is as follows: If the total surface area is kept constant, then it will be more accessible to reaction, the smoother it is. The opposite trends of the effect of fractal dimension on the reaction efficiency, under conditions of either fixed size or fixed area, are summarized in Fig. 13. Thus, for the optimization of the catalytic performance, both structural parameters are open to rational design.

5. Optimization of the Pore Size: Some Photophysical Observations

We report preliminary experimental observations obtained using photoprocess in the configuration used for the simulations of Section 4, namely:

$$A^* + Q \longrightarrow A \tag{13}$$

in which A^* is excited state of a molecule dissolved in a solution confined in pore space of porous materials, and Q is a quencher located on the surface at full coverage. Considerable interest in porosity effects on photoprocesses has been reported recently in several laboratories [42] including our own [43], but we are unaware of previous attempts to elucidate the pore-size effect in an Eley-Rideal mechanism in solution (i.e., below the Knudsen regime; for a study above this region, see [44]).

We report an observation of considerable pore size effect in the quenching reaction of the excited state of pyrene (Py, 1) by an iodinized surface, by means of a heavy atom mechanism. The heavy atom was introduced onto the surface of a series of controlled-pore glasses (cpg, pore size 150, 500 and 1000 Å) through routine synthetic procedures [45], resulting in the derivatized surface, 2. The quenching kinetics of Py excited at 337 nm with a PRA LN-1000 nitrogen laser was followed at 397 nm. The reported lifetime, τ, is the value of decay to e^{-1} of the maximal amplitude (although the fit to first order decay was rough; thus τ here is only a comparative measure). The τ_0 values are those recorded for the supernatant liquid.

The quenching of Py^* by iodine is not purely diffusion controlled (e.g., for ethyliodide in solution the quenching rate is $7 \cdot 10^7$ M^{-1} sec^{-1} [46]. However, a viscosity study, in which τ_0/τ was determined as a function of the viscosity of a series of alcohols which served as pore-filling solvents, clearly demonstrates (Fig. 14) the slowing of the reaction as viscosity increases, i.e., the diffusive nature of the transport and its Eley-Rideal nature. The main result, as seen in Fig. 14, is that for high efficiency of catalytic-activity (quenching, in our case) the narrow pore-size should be used. The pore size effect was especially pronounced for the lowest-viscosity solvent, (methanol), as is seen for the left-most points in Fig. 14.

Since the surface fractal dimension of these materials is low (2.2-2.3 [47]), the observed effect is attributed mainly to the pore size: Since Py concentration in all cases was kept constant ($1 \cdot 10^{-4}$M), the only difference between the various cpgs is in the average distance that a Py^* molecule is located from the quenching wall.

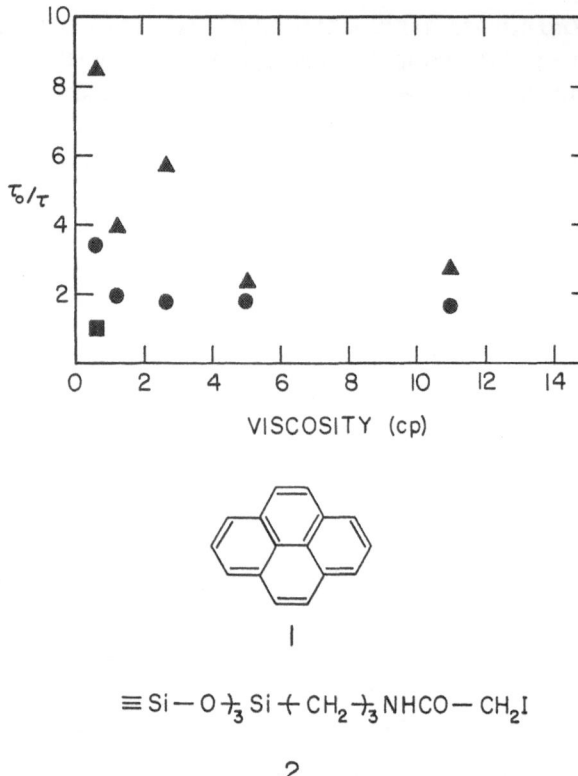

Fig. 14: Quenching of excited state pyrene (1), by iodine-derivatized surfaces (2) of controlled pore glass with average pore size 150 Å (▲), 500 Å (●) and 1000 Å (■). Pyrene was dissolved in alcohols of increasing viscosities, from left to right: methanol, ethanol, n-propanol, n-pentanol. Shown are the effects of pore size and of viscosity on pyrene life time (τ - within the pores; τ_0 - in the supernatant).

Indeed, (Fig. 14) for cpg-1000Å, $\tau_0/\tau=1$. Attempts to fit reaction 13 to equation (7)-type models are in progress, also taking into account the self-decay, τ_0.

Acknowledgments: Supported by the U.S.-Israel Binational Foundation, by the Belfer Foundation, by the Aronberg Foundation (to D.F.) and by the Berman Foundation (to O.C.). Discussions with R. Kosloff (Section 2) and with P. Pfeifer (Section 4) were very helpful and are gratefully acknowledged.

References

1.   P. Laszlo: *Acc. Chem. Res.* **19**, 121 (1986); P. Laszlo: In *Preparative Chemistry Using Supported Reagents* ed. by P. Laszlo, Chap. 1 (academic Press, San-Diego, 1987)

2. G.A. Somorjai: *Chemistry in Two Dimensions: Surfaces,* (Cornell University Press, Ithaca, 1981)

3. U. Müller, K.K. Unger: In *Characterization of Porous Solids,* ed. by K.K. Unger, J. Rouquerol, K.S.W. Sing, H. Kral (Elsevier, Amsterdam, 1988), p. 101

4. a) B.B. Mandelbrot: *The Fractal Geometry of Nature* (Freeman, San-Francisco, 1982); b) H. Takayasu: *Fractals* (Asakura-Shoten, Toyou, 1986)

5. J. Feder: *Fractals* (Plenum, New-York, 1988); H.E. Stanley, N. Ostrowski: ed's, *On Growth and Form* NATO ASI Ser. (Nijhoff, Dordrecht, 1986); L. Pietronero, E. Tossati: ed's, *Fractals in Physics* (North Holland, Amsterdam, 1986)

6. D. Avnir, ed.: *The Fractal Approach to Heterogeneous Chemistry* (Wiley, Chichester, 1989)

7. A. Le-Mehaute: Chap. 4.1.4 in ref. 6

8. R. Kopelman: *J. Stat. Phys.* **42,** 185 (1986); L.W. Anacker, R. Kopelman: *Phys. Rev. Lett.* **58,** 289 (1987); J. Prasad, R. Kopelman: *J. Phys. Chem.* **91,** 265 (1987)

9. G. Daccord, L. Lenormand: *Nature (London)* **325,** 41 (1987); G. Daccord: *Phys. Rev. Lett.* **58,** 479 (1987)

10. D. Farin, D. Avnir: *J. Phys. Chem.* **91,** 5517 (1987)

11. D.P.C. Fung, C. Fairbridge, R. Anderson: *Fuel,* **67,** 753 (1988)

12. E. Ignatzek, P.J. Plath, U. Hundorf: *Z. Phys. Chem. (Leipzig)* **268,** 859 (1987)

13. R.M. Ziff, K. Fichthorn: *Phys. Rev. B* **34,** 2038 (1986); K.A. Fichthorn, R.M. Ziff, E. Gulari: In *Catalysis 1987* ed. by J.W. Ward (Elsevier, Amsterdam, 1988), p. 883

14. a) W.-G. Ackerman, H. Spindler: *Z. Phys. Chem. (Leipzig)* **269,** 1000 (1988); b) N. Vlachopoulos, P. Liska, J. Augusinski, M. Gratzel: *J. Am. Chem. Soc.* **110,** 1216 (1988); c) C.-I. Yang, M.A. El-Sayed, S.L. Suib: *J. Phys. Chem.* **91,** 4440 (1987)

15. J.J. Carbery: *J. Catal.* **107,** 248 (1987); J.J. Carbery: *J. Catal.* **114,** 277 (1988)

16. P. Meakin: *Chem. Phys. Lett.* **123,** 428 (1986)

17. P. Pfeifer, D. Avnir, D. Farin: *J. Stat. Phys.* **36,** 699 (1984); **39,** 263 (1985)

18. P. Pfeifer: *Chimia* **39,** 120 (1985)

19. D. Farin, D. Avnir: In *Characterization of Porous Solids* ed. by K.K. Unger, D. Behrens, H. Kral, p. 421 (Elsevier, Amsterdam, 1988)

20. D. Farin, D. Avnir: *J. Am. Chem. Soc.* **110,** 2039 (1988)

21. a) D. Farin, D. Avnir: In *Proc. 9th Congress on Catalysis,* Vol. 3, p. 998, *Characterization and Metal Catalysts* ed. by M.J. Phillips, M. Ternan (Chem.

Inst. Canada, 1988) b) A. Seri-Levy, J. Samuel, D. Farin, D. Avnir: In *Photochemistry on Solid Surfaces* ed. by M. Anpo, T. Matsuura (Elsevier, Amsterdam, 1989), p. 353

22. T.H.P. Chang et al: *IBM J. Res. Develop.* **32**, 462 (1988)

23. C.B. Roxolo, H.W. Beck, B. Abeles: *Phys. Rev. Lett.* **57**, 2462 (1986)

24. G. Guo, X. Huang, Y. Tang: preprint, 1988

25. S. Aubry: *J. Phys.* **44**, 147 (1983)

26. T. Kleiser, M. Boceck: *Z. Metallkde* **77**, 582 (1986); B. Sprusil, F. Hnilica: *Czech. J. Phys.* **35**, 897 (1985)

27. R. Silva, O.L. Perez, A.E. Gonzalez, M.J. Yacaman: *Phys. Rev. Lett.* **57**, 2552 (1986)

28. T. Pajkossy, L. Nyikos: *Electrochim. Acta* **34**, 181 (1989)

29. M.L. Kagan, O. Citri, R. Kosloff, D. Avnir: *J. Phys. Chem.* **93**, 2728 (1989)

30. D. Kosloff, R. Kosloff: *J. Compt. Phys.* **52**, 35 (1983)

31. A. Seri-Levy, D. Farin, O. Citri, D. Avnir: *Proc. Klaus Symp. on Tribology, ACS Symp. Ser.* ed by S. Hsu et al (1989, in press)

32. O. Citri et al: in preparation

33. a) D. Avnir, D. Farin, P. Pfeifer: *J. Colloid Interface Sci.* **103**, 112 (1985) b) D. Avnir, D. Farin, P. Pfeifer: *Nature (London)* **308**, 261 (1984)

34. a) H. Van Damme, J.J. Fripiat: *J. Chem. Phys.* **82**, 2785 (1985) b) C. Fairbridge, H.S. Ng, A.D. Palmer: *Fuel* **65**, 1759 (1986) c) S.H. Ng, C. Fairbridge, B.H. Kaye: *Langmuir* **3**, 340 (1987) d) R.R. Mather: Preprint 1989 e) M. Ben Ohud, F. Obrecht, L. Gatineau, P. Levitz, H. Van Damme: *J. Coll. Interface Sci.* **156**, 124 (1988) f) H. Van Damme, P. Levitz, L. Gatineau, J.F. Alcover, J.J. Fripiat: *J. Colloid Interface Sci.* **122**, 1 (1988)

35. P. Pfeifer, D. Avnir: *J. Chem. Phys.* **79**, 3558 (1983); **80**, 4573 (1984)

36. M. Boudart, G. Djega-Mariadassou: *Kinetics of Heterogeneous Catalytic Reactions* (Princeton University Press, Princeton, 1982)

37. M. Silverberg, D. Farin, A. Ben-Shaul, D. Avnir: *Ann. Isr. Phys. Soc.* **8**, 451 (1986)

38. C.W. Curtis, J.A. Guin, B.L. Kamajian, B.L. Moody: *Fuel Process Technol.* **12**, 111 (1986)

39. L. Nyikos, T. Pajkossy: *Electrochim. Acta* **31**, 1347 (1986)

40. P.-G. de-Gennes: *C.R. Hebd. Sc. Acad. Sci. Ser. II* **295**, 685 (1982)

41. A. Seri-Levy, P. Pfeifer, D. Avnir: in preparation

42. N.J. Turro: *Tetrahedron* **43**, 1589 (1987)

43. E. Wellner, M. Ottolenghi, D. Avnir, D. Huppert: *Langmuir* **2**, 616 (1986)

44. J.M. Drake, P. Levitz, N.J. Turro, K.S. Nitche, K.F. Cassidy: *J. Phys. Chem.* **92**, 4680 (1988)

45. H.H. Weetal: In *Methods in Enzymology* ed. by K. Moosbach, Vol. 44 (Academic Press, New-York, 1976)

46. T. Medinger, F. Wilkinson: *Trans. Faraday Soc.* **62**, 1785 (1962)

47. D. Pines, D. Huppert: *Chem. Phys. Lett.* submitted, 1989; A. Hohr, H.-B. Neuman, M. Steiner, P.W. Schmidt, P. Pfeifer, D. Avnir: *Phys. Rev. B* **38**, 1462 (1988)

Reaction Kinetics in Disordered Systems: Hierarchical Models

G. Zumofen[1], *A. Blumen*[2], *and J. Klafter*[3]

[1]Laboratorium für Physikalische Chemie, ETH-Zentrum,
 CH-8092 Zürich, Switzerland
[2]Physics Institute and BIMF, University of Beyreuth,
 D-8580 Bayreuth, Fed. Rep. of Germany
[3]School of Chemistry, Tel-Aviv University, Tel-Aviv, 69978 Israel

Abstract. We investigate diffusion-limited reactions of the types A + A → 0 and
A + B → 0 in disordered systems by modelling the dynamics through random walks. By
introducing hierarchical structures, two aspects of disorder are considered: (a)
temporal and (b) energetic. The temporal disorder is accounted for by continuous
time random walks (CTRW), whose waiting times distribution displays long time tails.
The energetic disorder is modelled using ultrametric spaces (UMS) with hierarchically
distributed energy barriers. We discuss the interplay between these two disorder
aspects in terms of subordination.

1. Introduction

All chemical reactions require that the participating species be brought close enough
to allow for interactions that result in the actual reaction. Therefore, in many
cases, the diffusion of the reactants plays a vital role in the reaction process.
[1,2] The effects of the medium in which diffusion takes place, and the time
evolution of the reaction are of prime interest to physical chemists and chemical
physicists.

In spite of the importance of diffusion-limited reactions, it is only recently
that extensive work on the role of disorder and fluctuations in reaction kinetics has
become feasible. [3-7] This is due to the emergence of new approaches for treating
randomness analytically and due to the recent enhancement in computing capacity.
These investigations have revealed qualitative deviations from the accepted
Smoluchowski-type decay laws as a result of spatial fluctuations and of disorder
(randomness) in the medium.

Let us start by recalling the basic facts and the importance of randomness for
our problem. Thus, the transport in a spatially random system (e.g. a mixed crystal
or an alloy) is determined by the distribution of microscopic (site-to-site) transfer
rates (temporal disorder) and by the (in general stochastic) interactions with the
surroundings (energetic disorder). Taking into account such microscopic aspects is
a hard task, which calls for large-scale numerical calculations. One notes from the
beginning that there is little hope of dealing with the full complexity of the
problem in terms of perturbations from ideal, non-random situations. [8,9]

In this contribution, we focus on a <u>sui-generis</u> treatment for randomness and
present two classes of models for disorder; these classes have engendered widespread
interest only during the last decade. Here, the standard models can be viewed as
being connected to a particular aspect of randomness: continuous-time processes [10-
14] exemplify the temporal and ultrametric structures [15-18] the energetic disorder.
Fractals [19,20] which model spatial disorder will not be discussed here and those
readers interested in reaction on fractals are referred to [21-23]. More advanced
models connect several aspects of randomness, and we will touch on a few implications
in the course of this presentation.

Basic to our understanding of disorder is the temporal evolution of the systems
under study. A comparison of observed relaxation patterns reveals that the decays
which are observed in disordered materials seldom have kinetic chemical counterparts.
The fundamental reason underlying this observation is that the kinetic scheme
operates under the tacit assumption of a 'well-stirred' reactor. As a rule,

disordered media lack such homogeneity, a fact which leads to interesting deviations from the kinetic picture.

We start by presenting the standard chemical-kinetics scheme, for which we derive the relaxation behaviors. These decay forms are then confronted with the more complicated patterns, which are actually used to fit relaxation in random materials. We then introduce the random walk picture for reaction dynamics and outline the basic ideas behind continuous time random walks and ultrametric spaces. Finally, we center on bimolecular-kinetics of $A + A \to 0$ and $A + B \to 0$ type, which we model through random walks.

2. The Basic Kinetic Approach

In this section we remind the reader of the basic kinetic scheme. As will soon become apparent, the decay laws which follow from the kinetic formalism are inadequate for a description of the intriguing relaxation forms found in disordered systems.

General irreversible reactions have the form:

$$A_1 + A_2 + \ldots + A_n = \sum_{i=1}^{n} A_i \overset{k}{\to} 0 \tag{1}$$

to which, in the kinetic scheme, corresponds the system of (in general) nonlinear differential equations

$$\frac{dA_i(t)}{dt} = -k \prod_{i=1}^{n} A_i(t) . \tag{2}$$

Here, the $A_i(t)$ are the concentrations of the i-th molecular species, whose initial values we will denote by A_{i0}. The simplest case of (1) is the unimolecular reaction (n=1) $A \overset{k}{\to} P$. The solution of the corresponding equation (1), for n=1, is exponential:

$$A(t) = A_0 e^{-kt} . \tag{3}$$

On the other hand, bimolecular reactions (n=2) $A + B \overset{k}{\to} 0$ obey the following set of kinetic equations:

$$\frac{dA(t)}{dt} = -k A(t)B(t) = \frac{dB(t)}{dt} \tag{4}$$

Equation (4) is readily solved by observing that $C = B_0 - A_0$ is a constant of motion. Writing $B(t) = C - A(t)$ the system decouples and we obtain:

$$\frac{1 + C/A(t)}{1 + C/A_0} = e^{Ckt} . \tag{5}$$

From (5) we infer that for $B_0 \gg A_0$ one has $C \approx B_0$, and thus $C/A(t) \gg 1$. Hence:

$$A(t) \approx A_0 e^{-B_0 kt} . \tag{6}$$

It follows that the decay of the minority species is quasiexponential. On the other hand, if $A_0 = B_0$ then $C = 0$ in (5). An expansion in small C leads to the decay form

$$A(t) = \frac{A_0}{1 + A_0 kt} \tag{7}$$

from which at longer times, $t \gg (A_0 k)^{-1}$, an algebraic time dependence emerges:

$$A(t) \propto \frac{1}{kt} \tag{8}$$

A very similar behavior also obtains for the A + A → 0 reaction, which obeys the kinetic equation:

$$\frac{1}{2}\frac{dA(t)}{dt} = -k[A(t)]^2 .$$ (9)

Here, the separation of variables and subsequent integration lead to:

$$A(t) = \frac{A_0}{1 + 2A_0kt}$$ (10)

This expression is very similar to (7). Thus the long-time decay is given by:

$$A(t) \propto \frac{1}{2kt} .$$ (11)

From this analysis, it follows that unimolecular and bimolecular reactions lead in the kinetic scheme to exponential or to 1/t algebraic forms in the long-time regime. These simple temporal patterns are due to: (a) The absence of large disorder effects and (b) The basic assumption of the 'well-stirred reactor'. One should be aware of the fact that the use of (2) implies a homogeneous spatial distribution of particles during the whole course of the reaction. That such an assumption is in general untenable, was discussed by us in previous work; [7,21-23] in fact non-homogeneous conditions are widespread. Furthermore, even under homogeneous initial conditions, the microscopic reactions by themselves create non-homogeneities and enhance already existing density fluctuations. Diffusion can only partly wipe out such effects [3,6,7,22,23] and then only when the diffusion length is large compared to the mean interparticle distance. At low particle densities, diffusion (or stirring) cannot create a homogeneous background.

Let us now turn to experimentally observed decay forms. These expressions are, in general, more complex than (5) ff. Exemplarily, one often finds:

I) Stretched-exponential, Kohlrausch-Williams-Watts [24-26] laws:

$$\Phi(t) = \exp[-(t/\tau)^\alpha]$$ (12)

Such dependencies were found in amorphous Si:H [27] where the defect-annihilation by diffusing hydrogens was monitored. The same form is also present when observing abstraction reactions of hydrogen atoms from matrix molecules [28,29]. Recent work on the time evolution of solvation may also be described using (12). [30]

II) Exponential-logarithmic relaxation patterns (enhanced power laws):

$$\Phi(t) = \exp[-C\ell n^\beta(t/\tau)]$$ (13)

Such expressions are of considerable use in describing electron scavenging and electron-hole recombination; they also appear in the analysis of relaxation phenomena related to hole-burning in glasses. [31,32]

III) General algebraic decays

$$\Phi(t) \propto (t/\tau)^{-\xi}$$ (14)

as for instance reported for the relaxation processes of photogenerated carriers, which occur after electron-hole-pair creation in amorphous Si:H. [33] A different application of Eq. (14) concerns the effect of chromatographic tailing, in which dissimilar chemical species display different trapping-release patterns during their hindered diffusion through chromatographic columns. [34]

3. Random Walks

We now turn our attention to relaxation mechanisms which can be modelled by reaction dynamics schemes within the framework of random walk theories. [35-38] We begin with pseudo-unimolecular reactions (in which we have a minority and a majority species) and we focus on deviations of the relaxation pattern from exponentiality. In the simplest models, one has one A and several B particles, and the A particle is annihilated on encountering a B particle. Depending on which of the species performs the motion, one distinguishes between the trapping model [36-43] (only A moves), the target (scavenging) model [7,21-23,44-47] (only B moves) and the moving targets [48] (both species move). We monitor the survival probability of the A particle averaged over all possible realizations of particle distributions and motions.

For simplicity, we discuss only nearest-neighbor random walks and center on the target model which is exactly solvable both for regular lattices [47] and for ultrametric spaces. [49] In the target model, one considers an immobile A molecule which is annihilated by the mobile B species. At the start, the non-interacting B molecules are randomly distributed, with probability p, over the structure. For a finite, relatively large system of N_T sites, one has $p \approx B_0/N_T$. At each step, we move the B-molecules to neighboring sites, the different directions occurring with equal probability. The reaction is assumed to happen instantaneously, when the first B-molecule lands on the site occupied by the A-particle. The survival law obtains by averaging over all possible initial distributions and over all realizations of the random walks for the B-particles. Several possibilities may now be envisaged in creating the initial distribution of B particles; [21,47] for instance, one may take the Poisson distribution:

$$g(j) = e^{-p} \frac{p^j}{j!}$$

(15)

In (15), $g(j)$ is the normalized probability of having j B particles at one site when the average number density is p.

Let us now call $F_m(r)$ the probability that a random walker starting from r, reaches the origin O (assumed to be the site at which the A-particle is located) for the first time in the mth step. For regular lattices, because of the symmetry of the walk $F_m(r)$ is also the first-passage time from O to r, as defined by Montroll and Weiss. [50] Furthermore, we denote by $H_n(r)$ the probability that a first passage from r to O occurs in the first n steps and have:

$$H_n(r) \equiv \sum_{m=1}^{n} F_m(r)$$

(16)

The probability, therefore, that a walker from r did not reach O in the first n steps is thus

$$\Phi_n(r) = 1 - H_n(r)$$

(17)

Using (17) we obtain the survival probability of the A molecule by appropriately weighting products of the $\Phi_n(r)$ functions:

$$\Phi_n = \prod_r' \{ \sum_j g(j)[\Phi_n(r)]^j \} .$$

(18)

Here the product extends over all structure sites, with the exception of the origin. Inserting (15) and (17) into (18) leads to

$$\Phi_n = \prod_r' \left\{ \sum_j e^{-p} \frac{[p \, \Phi_n(r)]^j}{j!} \right\}$$

$$= \prod_r' \exp[-p + p\Phi_n(r)] = \exp[-p \sum_r' H_n(r)]$$

(19)

85

For regular lattices (19) may be further simplified since, according to equations (III.2) and (III.3) of [50], one has

$$\sum_r{}' F_m(\mathbf{r}) = S_m - S_{m-1} \qquad (m \geq 1) .$$

(20)

Here, S_m denotes the mean number of distinct sites visited by a random walker in m steps, with $S_0 = 1$. Equation (20) may also be derived directly, by noting that the increase of S_m is given by the <u>total</u> number of new sites visited in the mth step. Introducing (16) and (20) into (19), one has <u>exactly</u>

$$\Phi_n = \exp[-p(S_n-1)] .$$

(21)

For regular lattices and for a not too small n one now has: [36,51]

$$d = 1 : \quad S_n = a_1\sqrt{n} + a_2/\sqrt{n} + \ldots$$

(22)

$$d = 2 : \quad S_n = \frac{a_1 n}{\ell n(a_2 n)} + \ldots$$

(23)

$$d = 3 : \quad S_n = a_1 n + a_2\sqrt{n} + \ldots$$

(24)

where the a_i are constants which depend on the lattice structure.

We now note that the long-time behavior of (21) stays exponential in three dimensions, but becomes nonexponential for $d = 1$ and $d = 2$. It is interesting to see that for $d = 1$, a Williams-Watts relaxation pattern, (12) emerges, with $\alpha = \frac{1}{2}$. This result is identical to former findings for the target problem, which used the Glarum model of dipolar relaxation in glasses as an experimental application. [45,46,52-56] As long as only random walks on regular lattices are envisaged, it is not obvious how the target model should be extended to obtain non-trivial Williams-Watts forms with other α values ($\alpha \neq \frac{1}{2}$, $\alpha \neq 1$). [57-60] In the following we show that general α values are the rule when CTRW or ultrametric structures are considered.

4. Continuous Time Random Walks (CTRW).

In continuous-time processes one relaxes the condition assumed in the previous section that the steps of a walker can occur only at preassigned times. In the CTRW formalism, one introduces a waiting-time distribution $\psi(t)$, which gives the probability density that the time between steps equals t. The CTRW method has been rendered popular by a classic work by Montroll and Weiss, [50] who used it to include $\psi(t)$ in the generating-function formalisms for random walks. For generalizations including space-time correlated waiting-time distributions, see [61-63].

Let us begin by listing a few widely used $\psi(t)$ forms. The simplest situation obtains for a memoryless process; in such a case the probability of staying at a given site during the time t is exponential:

$$\Psi(t) = e^{-qt} .$$

(25)

If $\psi(t)$ is the probability density that an event occurs at a time t after the previous event has taken place, then obviously

$$\psi(t) = -\frac{d}{dt} \Psi(t)$$

(26)

and (25) leads to the Poisson process

$$\psi^p(t) = q \exp(-qt).$$

(27)

Slightly more complex forms obtain by using expressions due to multipolar or exchange interactions in the presence of substitutional disorder for $\psi(t)$. [64] These forms are well-behaved, in that for them, the first moment τ_1 is finite:

$$\tau_1 = \int_0^\infty dt\ t\ \psi(t) \tag{28}$$

Much more interesting are the distributions for which the integral in (28) is divergent. Thus Scher and Montroll [12] have modeled transport in amorphous media through a $\psi(t)$ which displays a long-time tail:

$$\psi(t) \propto t^{-1-\gamma} \qquad (0<\gamma<1) \tag{29}$$

Interestingly, such an expression is intimately related to a fractal set of event times. [14,65,66] From the Poisson process, in (27), one readily constructs a dilatationally symmetric distribution, by taking into account events on all time-scales, in the following way:

$$\psi(t) = \frac{1-N}{N} \sum_{j=1}^\infty N^j q^j \exp(-tq^j), \tag{30}$$

where $N < 1$. As is apparent, the distribution (30) is a normalized sum of Poisson terms and

$$\psi(qt) = \frac{\psi(t)}{(Nq)} - \frac{(1-N)}{N} \exp(-tq) \tag{31}$$

For later applications, we need $q < N$, so that $q < 1$ and thus at longer times $\psi(qt) = \psi(t)/Nq$. The last expression is equivalent to (29) when $\gamma = \ln N/\ln q$ is set. Equation (29) directly shows the temporal <u>scaling</u> of $\psi(t)$, i.e. its fractal nature in time.

We note that for algebraic $\psi(t)$ forms, scaling carries over to other quantities related to $\psi(t)$. Let $\chi_n(t)$ denote the probability that exactly n events occurred in time t. This basic quantity of the CTRW formalism is connected to $\psi(t)$ via its Laplace transform:

$$\mathcal{L}[\chi_n(t)] \equiv \chi_n(u) = [\psi(u)]^n \frac{[1-\psi(u)]}{u}, \tag{32}$$

where $\psi(u) = \mathcal{L}[\psi(t)]$.

As shown in [22], for $\psi(t)$ whose first moment τ_1 is infinite, the $\chi_n(t)$ coalesce at long times. For a given time t, the set of curves with n less than a certain parameter $n_{max}(t)$ scales, i.e. one has approximately

$$\chi_n(t) = \begin{cases} \chi_0(t) \propto t^{-\gamma} & \text{for } n < n_{max}(t) \\ 0 & \text{otherwise.} \end{cases} \tag{33}$$

On the other hand, for distributions whose first moments exist, scaling does not hold.

We now evaluate S(t), the mean number of sites visited by a walker in continuous time. Qualitatively, in situations in which $\tau_1 < \infty$, one expects the pattern to obey (22) to (24), with n replaced by t/τ_1. This result is indeed well fulfilled, as shown in extensive studies on (small) deviations due to higher order terms. [64] On the other hand, when τ_1 is infinite and $\psi(t)$ scales, t/τ_1 is meaningless. Then another argument may be used. From the additional temporal averaging required, S(t) is merely

$$S(t) = \sum_{n=0}^\infty S_n \chi_n(t). \tag{34}$$

We know now that, generally in the long-time domain, S_n has a power-law dependence on n. Setting $S_n \propto n^\epsilon$ and using (33) it follows that

$$S(t) \propto \sum_{n=0}^{n_{max}} n^{\varepsilon} \chi_0(t) \propto \chi_0(t) \, n_{max}^{1+\varepsilon} \quad . \tag{35}$$

Now the $\chi_n(t)$ are normalized, a fact which determines the time dependence of $n_{max}(t)$, [66]

$$1 = \sum_{n=0}^{\infty} \chi_n(t) \simeq \sum_{n=0}^{n_{max}} \chi_0(t) = \chi_0(t) n_{max} \quad . \tag{36}$$

For $\chi_0(t) \propto t^{-\gamma}$, one has $n_{max} \propto t^{\gamma}$ and hence, from (35), $S(t) \propto t^{\gamma\varepsilon}$. For dimensionality d, (where d denotes the Euclidean dimension for regular lattices and the spectral dimension for fractals) one obtains:

$$S(t) \propto \begin{cases} t^{\gamma d/2} & \text{for } d < 2 \\ t^{\gamma} & \text{for } d > 2. \end{cases} \tag{37}$$

In (37), the two exponents (γ for the temporal and $d/2$ for the spatial aspect) combine multiplicatively, i.e. the two aspects subordinate. [67]

We now consider the relaxation pattern of pseudo-unimolecular reactions $A + B \rightarrow 0$, $A_0 \ll B_0$ under continuous-time conditions. For the target problem introduced in Section 3 we have to extend the formalism to the CTRW situation. From (16) we now obtain the probability $H(t;r)$ of a visit from r to $\mathbf{0}$ in time t:

$$H(t;r) = \sum_{n=0}^{\infty} \chi_n(t) \, H_n(r) , \tag{38}$$

where the $\chi_n(t)$ are defined as in (32). Since, from (16) and (20) it follows that

$$\sum_{r}' H_n(r) = S_n - 1 , \tag{39}$$

in fact, using (34), one also has:

$$\sum_{r}' H(t;r) = S(t) - 1 . \tag{40}$$

Under the assumed independence of motion of the walkers in the target problem, the CTRW transformation of (18) and (19) carries through, and for the relaxation of the target, one obtains

$$\Phi(t) \propto \exp[-pS(t)] , \tag{41}$$

where $S(t)$ is given by (37). Equation (41) together with the subordinated $S(t)$ yield a stretched-exponential form. Now, the parameter α in Eq. (12) may take any value between 0 and 1, if one chooses γ accordingly, so that the CTRW provides an adequate extension for the Glarum model of relaxation in glasses, [52] as pointed out by Shlesinger and Montroll. [45,65]

5. Random Walks on Ultrametric Spaces

In general, sites in a disordered material are separated by energy barriers whose height is random. [23] Thus, a particle located on a certain site needs thermal energy in order to be able to surmount the surrounding barriers. A given activation energy only lets the particle visit a subset (cluster) of sites around the starting point. One may then classify the sites according to the energy required to reach them: [18] An ultrametric space (UMS) corresponds to such a classification. [17-19,23,68-70] As examples, Fig. 1 shows the regularly multifurcating UMS Z_2 and Z_3.

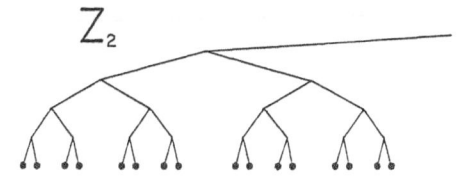

Z_2

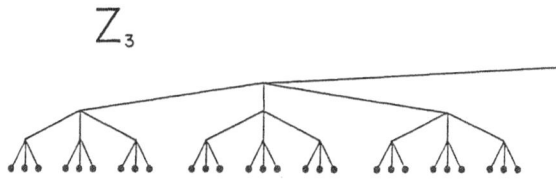

Z_3

Fig. 1.   The ultrametric spaces (UMS) Z_2 and Z_3.

The UMS Z_2 and Z_3 have branching ratios $b = 2$ and $b = 3$, and, for simplicity, we have assumed the barrier heights to be hierarchically arranged, so that all consecutive energy levels differ by Δ, the energies E being $E = m\Delta$, $m \in \mathbf{N}$. Note that only the points on the baseline belong to the UMS and that the structure above the baseline documents barrier heights and intersite connections. The height $d(x,y)$ of the barrier between sites x and y may be used as a generalized "distance". It may easily be verified that the reaction

$$d(x,y) \leq \max(d(x,w), d(y,w)) \tag{42}$$

holds for all UMS sites. Equation (42) is the "strong" triangle inequality which leads to the name *ultrametric*.

We now consider a simple qualitative argument for the temporal dependence of S(t), the mean number of sites visited during t, when the particles are thermally activated, so that the intersite transition rates are proportional to $R = \exp(-E/kT)$.

Let us focus on the time interval

$$e^{m\Delta/kT} \leq wt_m \leq e^{(m+1)\Delta/kT} . \tag{43}$$

During this time interval b^m points of the UMS Z_b are accessible to the walker, and one has

$$b^m \propto b^{(kT/\Delta)\ln(wt_m)} = e^{\delta \ln(wt_m)} = (wt_m)^\delta , \tag{44}$$

where we set

$$\delta = \frac{kT}{\Delta} \ln b . \tag{45}$$

For $\delta < 1$, b^m increases more slowly than t_m, and the walker explores practically all accessible points. This is the compact exploration case. Thus,

$$S(t) \simeq (wt)^\delta \propto t^\delta \quad (\delta < 1) . \tag{46}$$

On the other hand, for $\delta > 1$, b^m increases more rapidly than t_m, and the mean number of distinct sites visited stays proportional to t_m; hence

$$S(t) \propto t \quad (\delta > 1) . \tag{47}$$

From this simple qualitative argument we have obtained a result which parallels our findings for the CTRW, Equations (37), (46) and (47) can also be obtained rigorously, as shown in [49].

Following the procedure in Section 4, for translating number of steps n into time t and recalling that $\chi_n(t)$ is the probability of having performed exactly n steps during time t, in continuous time and for the probability of staying at the origin, [49] one has

$$P_0(t) = \sum_{n=0}^{\infty} \chi_n(t) \, P_{0,n} \tag{48}$$

which in Laplace-transformed form is

$$P_0(u) = \frac{1-\psi(u)}{u} \sum_{n=0}^{\infty} [\psi(u)]^n \, P_{0,n} \tag{49}$$

This enables us to address the issue of interplay between random walks on UMS and CTRW.

Now, following former work on UMS, [71-73] Bachas and Huberman [74] have succeeded in solving the master equation exactly for regularly multifurcating trees. They obtain:

$$P_0(t) = b^{-n} + (b-1) \sum_{m=1}^{n} b^{-m} \exp(-\lambda_m t/\tau) \tag{50}$$

where $P_i(0) = \delta_{i0}$ and

$$\lambda_m = (1-R/b) \sum_{p=m}^{n-1} R^p + R^n \tag{51}$$

with $R = \exp(-\Delta/kT)$. For b=2 and $\tau = 1/b$ the solution is that found in [72]. For an infinite tree, $n \to \infty$, (50) takes the form of a Weierstrass-series: [16,22,66]

$$P_0(t) = (b-1) \sum_{m=1}^{\infty} b^{-m} \exp[-CR^m(t/\tau)] \tag{52}$$

see (30), with q replaced by R and N by 1/b. In (52) $C = (b-R)/(b-bR)$. Note that $P_0(Rt) \propto bP_0(t)$, from which $P_0(t) \propto t^{-\delta}$ with $\delta = (\ln b)kT/\Delta$, i.e. (45) follows.

As discussed in [49], the solution (50) corresponds to the CTRW-process with a Poisson-distributed $\psi(t)$, (27). It then follows exactly that;

$$P_{0,n} = (b-1) \sum_{m=1}^{\infty} b^{-m} \exp(-nC'R^m) \tag{53}$$

with $C' = (b-R)/(bR-R)$. By Laplace-transformation of (52), one has

$$P_0(u) \propto \begin{cases} u^{\delta-1} & \text{for } \delta < 1 \\ \text{const} & \text{for } \delta > 1 \end{cases} \tag{54}$$

Furthermore, we are now in a position to consider CTRW with broad waiting-time distributions, as in (29). The starting point, as in [49], is the Laplace-transform of (52), together with (49). One obtains

$$P_0(t) \propto \begin{cases} t^{-\gamma d} & \text{for } \delta < 1 \\ t^{-\gamma} & \text{for } \delta > 1 \end{cases} \tag{55}$$

For $\gamma < 1$ and $\delta < 1$ the two coefficients combine multiplicatively in $P_0(t)$, i.e. the two processes subordinate. [67] It is also remarkable that for $\delta > 1$, the long-time behavior is dictated by the temporal disorder and that the energetic disorder is no longer important.

It can be shown that the same subordination holds for the mean number of distinct sites visited: [49]

$$S(t) \propto \begin{cases} t^{\gamma\delta} & \text{for } \delta < 1 \\ t^{\gamma} & \text{for } \delta > 1 \end{cases} \tag{56}$$

Using these results, we are now able to discuss relaxation phenomena on UMS. As in the previous sections, we restrict ourselves to the target problem, and refer the reader interested in the trapping problem to our former publications. [22,23,68-70]

Let us again start by noting that all sites of our regularly multifurcating UMS (fixed branching ratio b and equidistant barriers) are equivalent. We again denote the site on which the target sits as the origin, $r=0$, and assign integer number $r \in \mathbf{N}$, to the other sites by counting. Equations (16) to (19) hold also for UMS, when the position of the site ($\mathbf{r}$ in Section 3) is reinterpreted to be the ordinal integer r. Furthermore, (20) holds for UMS since $\Sigma_{\mathbf{r}} F_m(\mathbf{r}) \equiv \Sigma_r F_{r,m}$ gives the increase in the total number of visited sites in the mth step and thus equals $S_m - S_{m-1}$. Consequently, we rederive for UMS, as in (21), the exact decay law of A particles annihilated by the moving B species:

$$\Phi_n = \exp[-p(S_n-1)] \tag{57}$$

where the B molecules were initially Poisson distributed, see (15). From (56) one has:

$$\Phi_n \simeq \exp(-Cpt^{\eta}) \tag{58}$$

which for $\eta < 1$ shows a stretched-exponential form, (12).

In Figure 2, we present the decay in the target problem in the UMS $\mathbf{Z}_2$. The full lines give the decay obtained from S_n in conjunction with (57), whereas the dots indicate the direct simulation of the target annihilation. We have taken the density of the walkers to be $p = 0.2$, and the dynamics to take place over the UMS $\mathbf{Z}_2$. As expected, in every case, the agreement between the two forms is excellent. Note that depending on the temperature, one has a crossover from exponential decays for $\eta \simeq 1$, i.e. at higher temperatures (kT ln b > Δ), to stretched-exponential decays at lower temperatures (kT ln b < Δ).

We pause to note that these findings may be very general. The study of broader classes of hierarchical structures [75] lead to results similar to the ones which obtain for UMS.

To summarize, in this section we have analyzed pseudounimolecular relaxation behaviors for UMS - both for regular random walks and for CTRW. The findings parallel our previous results obtained for random walks and CTRW on regular lattices. This concludes our exposition on pseudo-unimolecular decays which obtain in disordered systems modelled through CTRW or ultrametric spaces. In the next section, we present an overview of bimolecular relaxation patterns.

6. The Bimolecular Reactions A + A → 0 and A + B → 0 (A$_0$ = B$_0$)

As stressed in Section 2, where we investigated the chemical kinetic scheme for the A + A → 0 and for the strictly bimolecular A + B → 0 (A$_0$ = B$_0$) reactions, in both cases under 'well-stirred' conditions, the long-time decay follows a 1/t law. Thus, (8) and (11), were the consequence of a spatially homogeneous situation. In this section, we determine the decay laws which apply in the presence of disorder, which will again be modelled through CTRW and UMS. The main type of relaxation pattern which will emerge from our studies is algebraic, $\Phi(t) \propto t^{-\xi}$, (14), where ξ may take any value between zero and one.

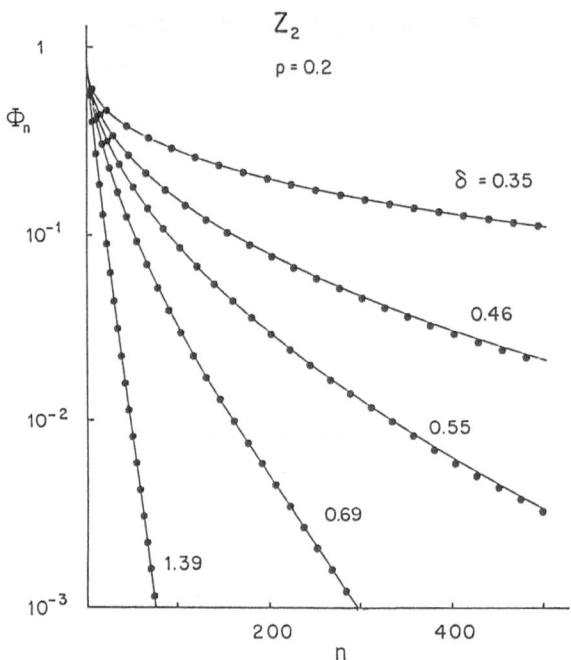

Fig. 2.   Decay law due to the annihilation of targets by random walkers on the UMS Z_2 . The concentration of walkers in $p = 0.2$ and the decays are monitored as a function of the temperature $\delta = (\ln 2)kT/\Delta$, for the values $\Delta/kT = 0.5$, 1, 1.25, 1.5 and 2. The full lines are the exact solution, (57), whereas the dots are the results of simulation calculations.

Here, we start with the $A + A \rightarrow 0$ reaction, since it will turn out to be <u>less</u> influenced by fluctuations than the strictly bimolecular $A + B \rightarrow 0$, $A_0 = B_0$, reaction. The $A + A \rightarrow 0$ reaction is of importance in energy transfer problems, where it describes exciton up-conversion and annihilation processes. [22,76-79] The study of pseudo-unimolecular reactions in the previous Sections has shown that the exponential law was modified by the change front to $S(t)$, [compare (6), (21), (41) and (57)]. Thus one may expect that the $A + A \rightarrow 0$ decay, which in the kinetic scheme is $1/t$, will follow a $1/S(t)$ law at longer times; [21,77] furthermore, one may suggest as an approximation to the decay,

$$\Phi_n^{AA} \simeq (1 + 2pS_n)^{-1} , \tag{59}$$

where S_n is given by (22 - 24). In [21] we have established by numerical simulations that (59) correctly describes the decay behavior on regular lattices (and on fractals). For the dynamical processes, periodic boundary conditions were used, so that the systems appeared infinite. The particles were placed randomly on the lattice, following a binomial (yes-no) distribution. At each step, the walkers moved one by one to neighboring sites. Any two particles that happened to occupy the same site during this process were immediately removed.

In Fig. 3 we display the results for the decay following the $A + A \rightarrow 0$ reaction on the linear chain, the square and the simple cubic lattice. The full lines give the simulation results obtained by averaging over 100 distinct initial conditions and walks each, whereas the dashed lines are the approximation, (59). We choose to plot $\ln \Phi_n^{AA}$ vs $\ln n$. In these scales, at long times the decays should turn into straight lines. This behavior is clearly apparent from the figure. Moreover, it so happens that (59) is almost quantitative in the range investigated. [21,22] An exact solution

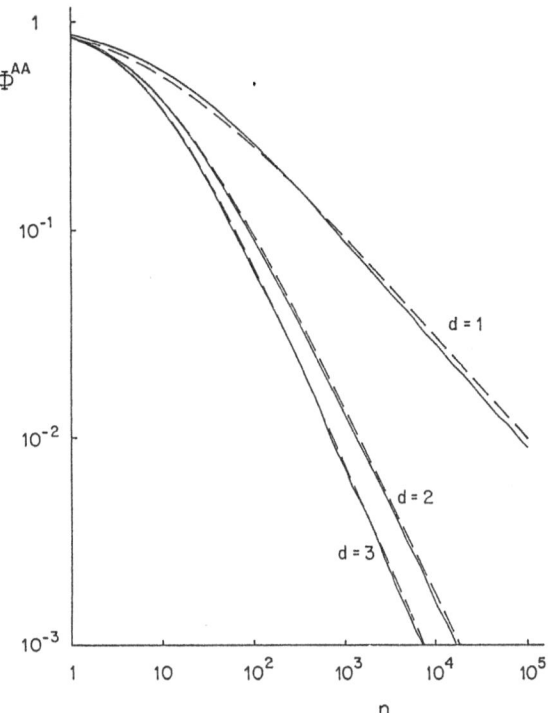

Fig. 3. Decays due to the bimolecular reaction A + A → 0. The full lines are the simulation results for the linear chain (d = 1), for the square (d =2), for the simple cubic lattice (d = 3). The dashed lines are Smoluchowski-type approximations, (59).

to the many-body problem involved in the A + A → 0 reaction was found by Torney and McConnell for d = 1 in the continuum, diffusion-equation limit. [79] In the range of Fig. 3, their expression is hardly distinguishable from the discrete, random-walk result. Hence, in all cases investigated, the long-time decay Φ_n^{AA} follows an algebraic form $\Phi_n^{AA} \propto n^{-\xi}$ (with $\xi = d/2$ for d < 2 and $\xi = 1$ for d > 2). Heuristically, one may view (59), the solution of a many-body problem, as being related to the probability of encounter of two particles, which itself is expressible, via S_n , by the volume visited by each.

We now turn to the A + A → 0 reaction under CTRW-conditions. The previous discussion for regular lattices has led to

$$\Phi_n^{AA}(t) \propto \begin{cases} n^{-d/2} & \text{for } d < 2 \\ n^{-1} & \text{for } d > 2. \end{cases} \tag{60}$$

If we view S_n as a measure of explored volume, it is tempting to envisage that in the CTRW scheme, S_n gets replaced in (59) by S(t), and therefore, for CTRW one should find:

$$\Phi^{AA}(t) \propto \begin{cases} t^{-\gamma d/2} & \text{for } d < 2, \gamma < 1 \\ t^{-\gamma} & \text{for } d > 2, \gamma < 1. \end{cases} \tag{61}$$

Equation (61) is then another example of subordination. [19,67]

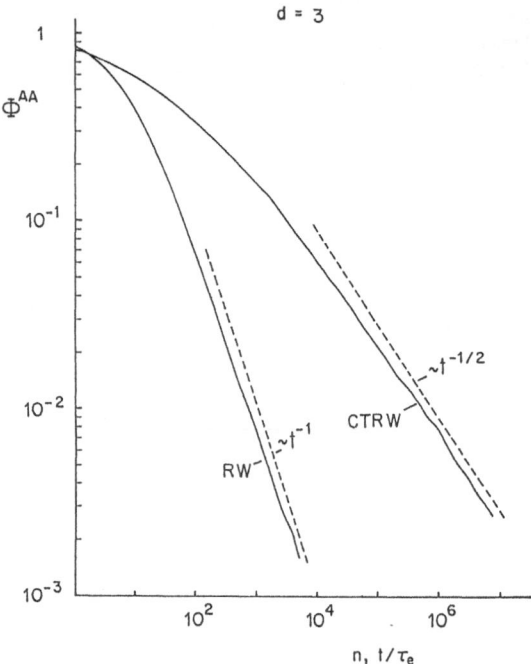

Fig. 4.   Decay laws Φ^{AA} for the A + A → 0 reaction on a simple cubic lattice, both for simple RW and for CTRW with $\gamma = 1/2$. The initial particle density is p = 0.1. The full lines are the simulation results averaged over 100 runs. The theoretical long-time slopes are indicated by dashed lines.

Our numerical simulations support (61) well. [66] In Fig. 4, we present the decay of the A + A → 0 reaction under CTRW conditions. We start from walkers on a simple cubic lattice, d = 3, and for $\psi(t)$ use a form with $\gamma = 1/2$. In the figure we present the corresponding decay law and contrast it with the simple RW results. Whereas at longer times, the random walk decay follows $1/t$, for the CTRW we find a $t^{-1/2}$ dependence at longer times, as may be verified by inspection of Fig. 4, in which these asymptotic slopes are also indicated.

To conclude our study of the A + A → 0 reaction, we also consider the influence of energetic randomness modelled through ultrametric spaces (UMS). At longer times, paralleling the previous discussion we expect, the relaxation pattern

$$\Phi^{AA}(t) \propto \begin{cases} t^{-\delta} & \text{for } \delta = (kT/\Delta) \ln b < 1 \\ t^{-1} & \text{else.} \end{cases} \tag{62}$$

The results of a typical calculation are presented in Fig. 5. The reaction depicted again takes place on the UMS Z_2 at a somewhat low temperature, so that $\delta = 0.347$. The initial density of particles is $p_A = 0.2$. The overall decay of Φ_n^{AA} for longer times (larger n) is quite well described by $1/n^{0.35}$, as indicated by the slope parallel to the decay. Thus, both on UMS and on regular lattices, Φ_n^{AA} may be well approximated by $\Phi_n^{AA} \propto S_n^{-1}$.

However, superimposed on this algebraic decay one finds slight fluctuations. To present them in a magnified form we have plotted the derivative $d(\ln \Phi^{AA})/d(\ln n)$ in the lower part of Fig. 5. The dots are the numerically evaluated slopes taken from the simulation, and the full line is the result of a smoothing procedure. The arrows indicate the estimates for the locations of the minima determined by a detailed

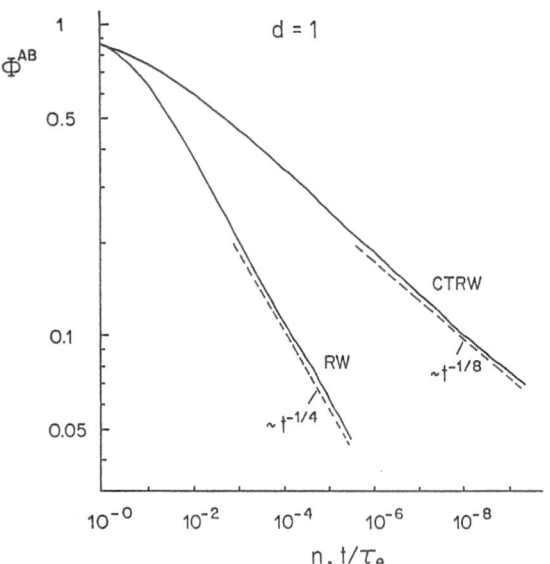

6. Decay law $\Phi^{AB}(t)$ for the strictly bimolecular $A + B \rightarrow 0$, $A(0) = B(0)$, with particles moving according to CTRW for a $\psi(t)$ with $\gamma = 1/2$ on the linear chain. The long-time slopes are also indicated.

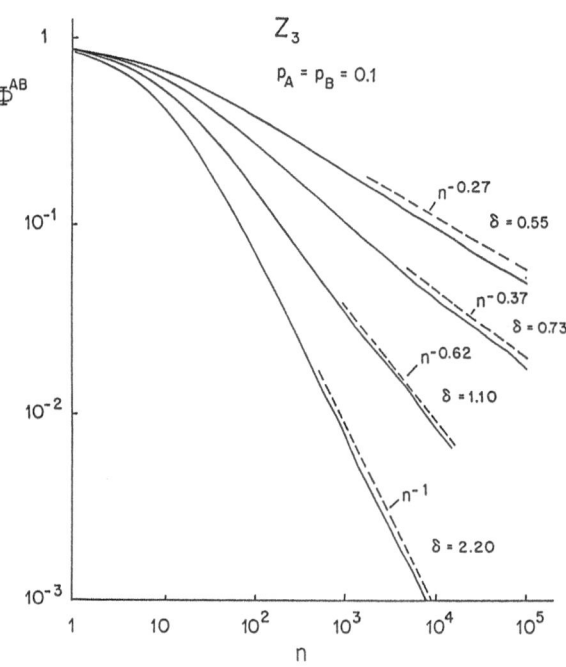

7. Decay pattern of the $A + B \rightarrow 0$, $A_0 = B_0$ reaction on the UMS Z_3. The initial number densities are $p_A = p_B = 0.1$. The temperature parameter Δ/kT varies from 0.55 to 2.2. The slopes for large n are indicated by dashed lines. Simulation calculations [66] for particles moving on a linear chain under the influence of a waiting-time distribution with $\gamma = 1/2$, support the conjectured subordination displayed by (64) as shown in Fig. 6.

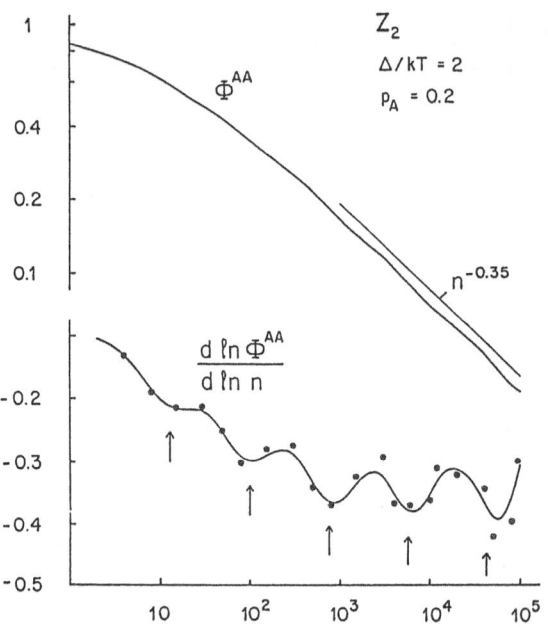

Fig

Fig. 5. Decay pattern for the A + A → 0 reaction on the UMS Z_2. Th
shows the decay for $\Delta/kT = 2$ and for an initial particle de
The slope of the decay for n large is also shown. In the low
variations of the slopes are given by dots, the full line i
a smoothing procedure. The arrows mark theoretically predi

analysis; in fact the minima may again be related to the cluster st
Increasing the temperature so as to have $\delta > 1$, again leads to a
expected from (62). As before, an increase in temperature leads one
result.

We now turn our attention to the strictly bimolecular reaction A +
We note from the start that the simplicity of the A + A → 0 reaction
over; the long-time regime is different from that of (59) to (62).
results from spatial fluctuations, which are <u>enhanced</u> by the chem
reaction. The reason for this effect is that at longer times, due to
the reaction, large regions containing only A or only B molecules a
diffusion no longer provides efficient stirring, and the reaction
slowly since mainly just the molecules at the boundaries of the A an
prone to react. The decay at longer times then follows: [4,6,7,21-2?

$$\Phi_n^{AB} \propto \begin{cases} n^{-d/4} & \text{for } d < 4 \\ n^{-1} & \text{for } d > 4. \end{cases}$$

The marginal dimension for the A + B → 0, $A_0 = B_0$ reaction is thus f(
When a broad waiting time distribution, (29), is superimpose(
subordination is again obtained through the time variable of $\psi(t)$,

$$\Phi^{AB}(t) \propto t^{-\gamma d/4} \quad \text{(for } d < 4, \gamma < 1).$$

We now address the question of the A + B → 0, $A_0 = B_0$ reaction o(
our previous knowledge of this reaction, we expect

$$\Phi^{AB}(t) \propto \begin{cases} t^{-\delta/2} & \text{for } \delta < 1 \\ t^{-1} & \text{for } \delta > 2 \end{cases}$$

Fi(

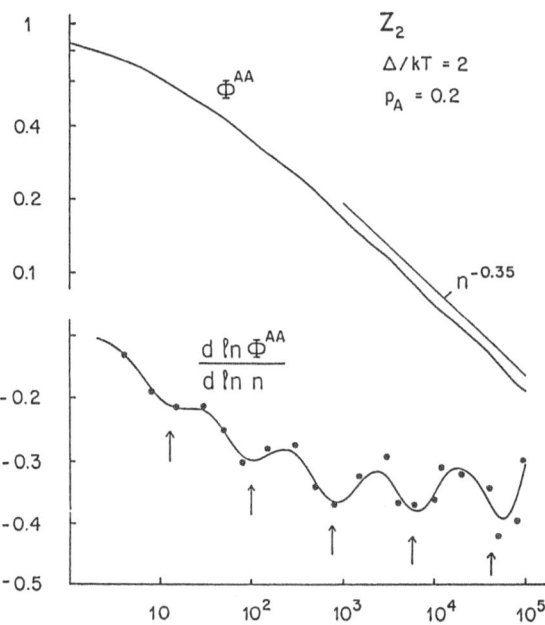

Fig. 5.   Decay pattern for the A + A → 0 reaction on the UMS Z_2 . The upper section shows the decay for $\Delta/kT = 2$ and for an initial particle density $p_A = 0.2$. The slope of the decay for n large is also shown. In the lower section, the variations of the slopes are given by dots, the full line is the result of a smoothing procedure. The arrows mark theoretically predicted minima.

analysis; in fact the minima may again be related to the cluster structure of Z_2 . Increasing the temperature so as to have $\delta > 1$, again leads to a 1/t pattern, as expected from (62). As before, an increase in temperature leads one to the kinetic result.

    We now turn our attention to the strictly bimolecular reaction A + B → 0, $A_0 = B_0$. We note from the start that the simplicity of the A + A → 0 reaction does not carry over; the long-time regime is different from that of (59) to (62). The difference results from spatial fluctuations, which are _enhanced_ by the chemical A + B → 0 reaction. The reason for this effect is that at longer times, due to the progress of the reaction, large regions containing only A or only B molecules appear. Then the diffusion no longer provides efficient stirring, and the reactions proceed more slowly since mainly just the molecules at the boundaries of the A and B regions are prone to react. The decay at longer times then follows: [4,6,7,21-23]

$$\Phi_n^{AB} \propto \begin{cases} n^{-d/4} & \text{for } d < 4 \\ n^{-1} & \text{for } d > 4. \end{cases} \tag{63}$$

The marginal dimension for the A + B → 0, $A_0 = B_0$ reaction is thus four. [4,6]
    When a broad waiting time distribution, (29), is superimposed on the walk, subordination is again obtained through the time variable of $\psi(t)$,

$$\tilde{\Phi}^{AB}(t) \propto t^{-\gamma d/4} \qquad \text{(for } d < 4, \ \gamma < 1). \tag{64}$$

    We now address the question of the A + B → 0, $A_0 = B_0$ reaction on UMS. Based on our previous knowledge of this reaction, we expect

$$\Phi^{AB}(t) \propto \begin{cases} t^{-\delta/2} & \text{for } \delta < 1 \\ t^{-1} & \text{for } \delta > 2 \end{cases} \tag{65}$$

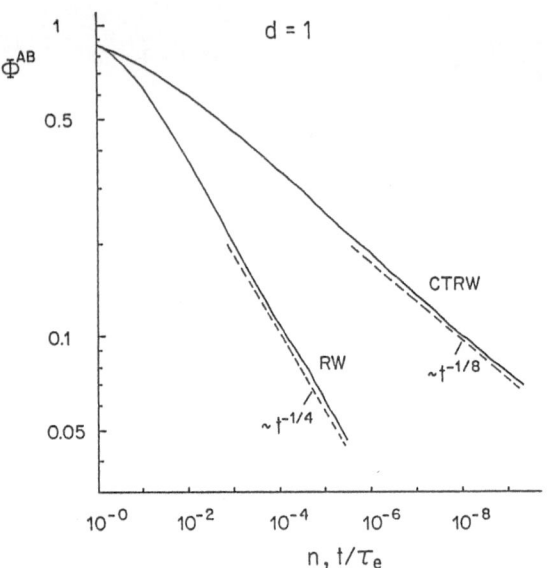

Fig. 6.   Decay law $\Phi^{AB}(t)$ for the strictly bimolecular $A + B \rightarrow 0$, $A(0) = B(0)$, with particles moving according to CTRW for a $\psi(t)$ with $\gamma = 1/2$ on the linear chain. The long-time slopes are also indicated.

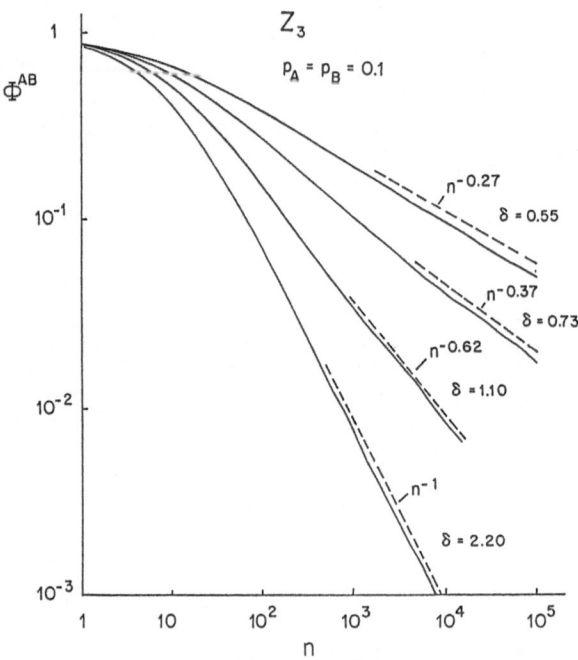

Fig. 7.   Decay pattern of the $A + B \rightarrow 0$, $A_0 = B_0$ reaction on the UMS Z_3. The initial number densities are $p_A = p_B = 0.1$. The temperature parameter Δ/kT varies from 0.55 to 2.2. The slopes for large n are indicated by dashed lines. Simulation calculations [66] for particles moving on a linear chain under the influence of a waiting-time distribution with $\gamma = 1/2$, support the conjectured subordination displayed by (64) as shown in Fig. 6.

with a crossover region for δ between 1 and 2. Simulation calculations [23] indeed show that (65) offers a very reasonable description of the long-time behavior for δ < 1 and for δ > 2.

In Fig. 7, we display the decay pattern Φ_n^{AB} which obtains for Z_3 and $p_A = p_B = 0.1$. [23] Here we average over five different realizations of initial conditions and walks. We take Δ/kT to be 0.5; 1; 1.5 and 2 so that δ varies between 0.55 and 2.2.

To monitor the algebraic forms we plot $\ln \Phi^{AB}$ vs. $\ln n$ in logarithmic scales. In Fig. 7 we have indicated by dashed lines the best fitted slopes to the decays. It is now apparent by inspection that for $\delta = 0.55$ and for $\delta = 0.73$, the fitted exponent to the algebraic decay is around $\delta/2$, i.e. δ is only one <u>half</u> of the value which holds for the A + A → 0 reaction. On the other hand, for quite large δ, such as δ = 2.2, we again recover the 1/t kinetic form.

The findings of Fig. 7 can also be explained by the fact that in the strictly bimolecular scheme, large regions containing only A- or only B-molecules appear in the course of the reaction. For long times, the decay proceeds more slowly, since only molecules at the boundary between the A- and the B-regions can react. We document this for UMS in Fig. 8. In this Figure, we present the course of a A + B → 0, $A_0 = B_0$ reaction on Z_3, where the five patterns are snapshots after n = 1, 50, $9 \cdot 10^2$, $3 \cdot 10^4$ and $2 \cdot 10^5$ steps. The UMS sites are arranged horizontally; a fifth of the Z_3 structure at the 12th hierarchical stage is shown in Fig. 7. Initially, the A and B particles are randomly distributed with $p_A = p_B = 0.1$. The course of the reaction magnifies the fluctuations; from an initially homogeneous situation, at later stages the A- and B-domains are well separated.

To conclude this section devoted to bimolecular reactions on regular lattices and on UMS under continuous-time conditions, we note that several microscopic models of a bimolecular type lead to the power-law forms, (59) to (65). In general, such decays may be well distinguished from other relaxation behaviors, such as stretched exponentials, by a sufficiently large dynamical range of measurements. [32] However,

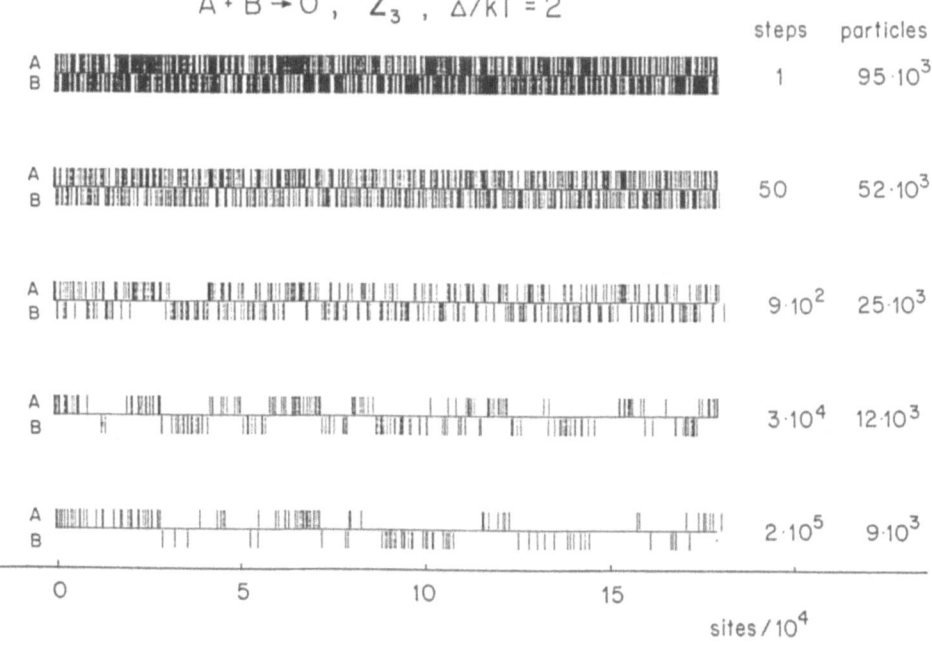

Fig. 8. Evolution of the A + B → 0, $A_0 = B_0$ reaction on the UMS Z_3. The snapshots are taken after the number of steps n indicated. The UMS is arranged horizontally and the A and B positions are indicated as up and down dashes, respectively. One-fifth of the Z_3 structure is shown at the twelfth stage.

distinguishing between different power-law decays may not be easy. Thus, additional experimental information such as concentration and temperature dependence may be necessary in order to pinpoint the microscopic relaxation behavior of a specific material displaying a power-law form.

7. Conclusions

As we have shown, deviations from 'well-stirred' relaxation behaviors are widespread, and extend from stretched-exponential to algebraic decays. Moreover, one may readily obtain such decay forms from theoretical approaches which include randomness. In the cases we considered (pseudo-unimolecular and bimolecular reactions), we were able to display the decay laws either in an analytically closed form or as a result of computer simulations. Disorder helps to differentiate between several reaction schemes, whose decays are similar under 'well-stirred' conditions. Thus the target problem, in which the minority species is stationary, has a more rapid decay than the trapping problem, in which only the minority species moves. The $A + A \rightarrow 0$ reaction has a marginal dimension of 2, whereas the marginal dimension of the $A + B \rightarrow 0$, $A_0 = B_0$ bimolecular reaction is 4, when the reactants are initially uncorrelated.

Underlying the basic models for disorder, there exist scaling symmetries, which lead to unexpected connections. Thus the δ parameter for a random walk on a UMS, involving temperature, activation energies and branching ratios, is similar in its effects to the parameter γ, which is given by the waiting-time distribution in the CTRW picture. Furthermore, CTRW processes may be applied to UMS; generally the CTRW parameter γ and the UMS parameter δ combine multiplicatively in the relaxation patterns: the processes subordinate.

Acknowledgement

The support of the Deutsche Forschungsgemeinschaft (SFB 213) and of the Fonds der Chemischen Industrie and grants of computer time from the ETH Rechenzentrum are gratefully acknowledged. J.K. acknowledges the support of the Fund for Basic Research administrated by the Israel Academy of Sciences and Humanities.

References

1. S. Chandrasekhar, Rev. Mod. Phys. 15, 1 (1943) and ref. mentioned, such as M. von Smoluchowski, Phys. Z. 17, 557, 585 (1916); Z. Phys. Chem. (Leipzig) 92, 129 (1917).
2. D.F. Calef and J.M. Deutch, Ann. Rev. Phys. Chem. 34, 493 (1983).
3. B. Ya. Balagurov and V.G. Vaks, Zh. Exp. Theor. Fiz. 65, 1939 (1973). [English translation Sov. Phys. JETP. 38, 968 (1974)].
4. A.A. Ovchinnikov and Ya. B. Zeldovich, Chem. Phys. 28, 215 (1978).
5. M.D. Donsker and S.R.S. Varadhan, Commun. Pure Appl. Math. 28, 525 (1975); 32, 721 (1979).
6. D. Toussaint and F. Wilczek, J. Chem. Phys. 78, 2642 (1983).
7. G. Zumofen, A. Blumen, and J. Klafter, J. Chem. Phys. 82, 3198 (1985).
8. P.W. Anderson in 'Ill-Condensed Matter' (eds R. Balian, R. Maynard and G. Toulouse), North Holland, Amsterdam (1979), p. 162.
9. T.S. Kuhn, 'The Structure of Scientific Revolutions' (2nd edn), Univ. of Chicago Press, Chicago (1970).
10. H. Scher and M. Lax, Phys. Rev. B7, 4491 (1973).
11. H. Scher and M. Lax, Phys. Rev. B7, 4502 (1973).
12. H. Scher and E.W. Montroll, Phys. Rev. B12, 2455 (1975).
13. G. Pfister and H. Scher, Adv. Phys. 27, 747 (1978).
14. E.W. Montroll and M.F. Shlesinger in 'Nonequilibrium Phenomena II: From Stochastics to Hydrodynamics' (eds J.L. Lebowitz and E.W. Montroll), North Holland, Amsterdam (1984).
15. N. Bourbaki, 'Eléments de mathématique, Topologie générale', Chap. IX, CCLS, Paris (1974).

16. A.D. Gordon, 'Classification', Chapman and Hall, London (1981).
17. W.H. Schikhof, 'Ultrametric Calculus', Cambridge Univ. Press (1984).
18. R. Rammal, G. Toulouse, and M.A. Virasoro, Rev. Modern Phys. $\underline{58}$, 765 (1986)
19. B.B. Mandelbrot, 'The Fractal Geometry of Nature', Freeman, San Francisco (1982).
20. K.J. Falconer, 'The Geometry of Fractal Sets', Cambridge Univ. Press (1985).
21. A. Blumen, G. Zumofen, and J. Klafter in 'Structure and Dynamics of Molecular Systems' (eds R. Daudel et al.), Reidel, Dordrecht (1985), p. 71.
22. A. Blumen, J. Klafter, and G. Zumofen in 'Optical Spectroscopy of Glasses', (ed. I. Zschokke), Reidel, Dordrecht (1986), p. 199.
23. A. Blumen, G. Zumofen, and J. Klafter, Ber. Bunsenges. Phys. Chem. $\underline{90}$, 1048 (1986).
24. R. Kohlrausch, Ann. Phys. (Leipzig) $\underline{12}$, 393 (1847).
25. G. Williams and D.C. Watts, Trans. Faraday Soc. $\underline{66}$, 80 (1970).
26. G. Williams, Adv. Polymer Sci. $\underline{33}$, 59 (1979).
27. J. Kakalios, R.A. Street and W.B. Jackson, Phys. Rev. Lett. $\underline{59}$, 1037 (1987).
28. V.L. Vyazovkin, B.V. Bol'shakov, and V.A. Tolkatchev, Chem. Phys. $\underline{75}$, 11 (1983).
29. T. Doba, K.U. Ingold, and W. Siebrand, Chem. Phys. Lett. $\underline{103}$, 339 (1984).
30. D. Huppert, V. Ittah, A. Masad and E.M. Kosower, Chem. Phys. Lett. $\underline{150}$, 349 (1988).
31. J. Friedrich and D. Haarer, Angew. Chem. Int. Ed. Engl. $\underline{23}$, 113 (1984).
32. J. Friedrich and A. Blumen, Phys. Rev. $\underline{B32}$, 1434 (1985).
33. J. Tauc, Semicond. Semimet. $\underline{21B}$, 299 (1984).
34. G.H. Weiss, Sep. Sci. Techn. $\underline{17}$, 1609 (1982-83).
35. M.N. Barber and B.W. Ninham, 'Random and Restricted Walks', Gordon and Breach, New York (1970).
36. G.H. Weiss and R.J. Rubin, 'Random Walks: Theory and Selected Applications', Adv. Chem. Phys., $\underline{52}$, 363 (1983).
37. Papers presented at the Symposium on Random Walks, J. Stat. Phys. $\underline{30}$, No. 2 (1983).
38. M.F. Shlesinger and B.J. West (eds), 'Random Walks and their Applications in the Physical and Biological Sciences', Amer. Inst. Phys., New York (1984).
39. G. Zumofen and A. Blumen, Chem. Phys. Lett. $\underline{88}$, 63 (1982).
40. H.E. Stanley, K. Kang, S. Redner, and R.L. Blumberg, Phys. Rev. Lett. $\underline{51}$, 1223 (1983).
41. K. Kang and S. Redner, Phys. Rev. Lett. $\underline{52}$, 955 (1984).
42. R. Kopelman, P.W. Klymko, J.S. Newhouse, and L.W. Anacker, Phys. Rev. $\underline{B29}$, 3747 (1984).
43. J. Klafter, A. Blumen, and G. Zumofen, J. Phys. Chem. $\underline{87}$, 191 (1983).
44. E.W. Montroll and J.T. Bendler, J. Stat. Phys. $\underline{34}$, 129 (1984).
45. M.F. Shlesinger and E.W. Montroll, Proc. Natl. Acad. Sci. USA $\underline{81}$, 1280 (1984).
46. S. Redner and K. Kang, J. Phys. $\underline{A17}$, L451 (1984).
47. A. Blumen, G. Zumofen, and J. Klafter, Phys. Rev. $\underline{B30}$, 5379 (1984).
48. A. Blumen, G. Zumofen, and J. Klafter, J. Physique Colloque, $\underline{Tome\ 46}$, C7-3 (1985).
49. G. Köhler and A. Blumen, J. Phys. $\underline{A20}$, 5627 (1987).
50. E.W. Montroll and G.H. Weiss, J. Math. Phys. $\underline{6}$, 167 (1965).
51. G. Zumofen and A. Blumen, J. Chem. Phys. $\underline{76}$, 3713 (1982).
52. S.H. Glarum, J. Chem. Phys. $\underline{33}$, 639 (1960).
53. P. Bordewijk, Chem. Phys. Lett. $\underline{32}$, 592 (1975).
54. J.E. Shore and R. Zwanzig, J. Chem. Phys. $\underline{63}$, 5445 (1975).
55. J.L. Skinner, J. Chem. Phys. $\underline{79}$, 1955 (1983).
56. A.A. Jones, P.T. Inglefield, J.F. O'Gara and A.K. Roy, in Transport and Relaxation in Random Materials, Ed. J. Klafter, R.J. Rubin and M.F. Shlesinger (World Scientific, Singapore, 1986).
57. A.A. Jones, in Molecular Dynamics in Restricted Geometries, Ed. J. Klafter and J.M. Drake (John Wiley, New York, 1989).
58. J.T. Bendler, D.G. Le Grand and W.V. Olszewski, in Transport and Relaxation in Random Materials, Ed. J. Klafter, R.J. Rubin and M.F. Shlesinger (World Scientific, Singapore, 1986).
59. G.W. Scherer (paper submitted to the International Congress on Glass, Leningrad, USSR, July 1989).

60. W.B. Jackson and J. Kakalios, in Advances in Amorphous Semiconductors I. Amorphous Silicon and Related Materials, Ed. H. Fritzche (World Scientific, Singapore, 1986).
61. J. Klafter and R. Silbey, Phys. Rev. Lett. $\underline{44}$, 55 (1980).
62. M.F. Shlesinger and J. Klafter, Phys. Rev. Lett. $\underline{54}$, 2551 (1985).
63. J. Klafter, A. Blumen and M.F. Shlesinger, Phys. Rev. $\underline{A35}$, 3081 (1987).
64. A. Blumen and G. Zumofen, J. Chem. Phys. $\underline{77}$, 5127 (1982).
65. M.F. Shlesinger, J. Stat. Phys. $\underline{36}$, 639 (1984).
66. A. Blumen, J. Klafter, and G. Zumofen in 'Fractals in Physics' (eds L. Pietronero and E. Tossatti), North Holland, Amsterdam (1986), p. 399; for applications of the Weierstrass form see: D. Haarer and A. Blumen, Angew. Chem. $\underline{100}$, 1252 (1988); H. Schnörer, D. Haarer, and A. Blumen, Phys. Rev. $\underline{B38}$, 8097 (1988).
67. A. Blumen, J. Klafter, B.S. White, and G. Zumofen, Phys. Rev. Lett. $\underline{53}$, 1301 (1984).
68. G. Zumofen, A. Blumen, and J. Klafter, J. Chem. Phys. $\underline{84}$, 6679 (1986).
69. A. Blumen, J. Klafter, and G. Zumofen, J. Phys. $\underline{A19}$, L77 (1986).
70. A. Blumen, G. Zumofen, and J. Klafter, J. Phys. $\underline{A19}$, L861 (1986).
71. S. Grossmann, F. Wegner, and K.H. Hoffmann, J. Physique Lett. $\underline{46}$, L575 (1985).
72. A.T. Ogielski and D.L. Stein, Phys. Rev. Lett. $\underline{55}$, 1634 (1985).
73. B.A. Huberman and M. Kerszberg, J. Phys. A18, $\underline{L331}$ (1985).
74. C.P. Bachas and B.A. Huberman, Phys. Rev. Lett. $\underline{57}$, 1965 (1986).
75. G.H. Köhler, E.W. Knapp, and A. Blumen, Phys. Rev. $\underline{B38}$, 6774 (1988).
76. P.W. Klymko and R. Kopelman, J. Phys. Chem. $\underline{87}$, 4565 (1983).
77. P. Argyrakis and R. Kopelman, J. Chem. Phys. $\underline{83}$, 3099 (1985) and references therein.
78. D.C. Torney and H.M. McConnell, Proc. Roy. Soc. London $\underline{A387}$, 147 (1983).
79. D.C. Torney and H.M. McConnell, J. Phys. Chem. $\underline{87}$, 1941 (1983).

Optimization and Complexity in Molecular Biology and Physics

P. Schuster

Institut für Theoretische Chemie der Universität Wien,
Währingerstr. 17, A-1090 Wien, Austria

Abstract. Many optimization problems in physics and biology lead to highly *rugged* cost functions. The most important examples come from the theory of spin glasses and from biological evolution. During the last decade new algorithms were developed in order to deal with such complex systems. Two techniques – known as simulated annealing and genetic algorithms – are applied to study the optimization of polynucleotide replication. Ruggedness of value landscapes in evolutionary optimization has its origin in the complexity of genotype-phenotype-relations. Folding of RNA molecules into spatial structures represents an example of a complex relation which is particularly important in virology and cell metabolism. A model of polynucleotide folding and optimization of replication rates through mutation and selection is presented and discussed. It enables us to extend the concepts of conventional population genetics into the realm of frequent mutations.

1. *Rugged* Cost Functions and Value Landscapes

During the last decade, more and more problems from mathematics, physics, biology and other scientific disciplines were found to be especially difficult to solve even with the assistance of powerful computers. Moreover, these problems have the unfortunate property that numerical efforts to find a solution increase rapidly, presumably exponentially, with the system size. According to this and other features which are used for classification purposes [1,2] and which will not be discussed here, these problems are characterized as *N(Nondeterministic) P(Polynomial time)-complete*. Problem solving techniques from physics and biology were generalized and applied to optimization tasks in other disciplines [3]. Three classes of methods proved particularly important:

- *Simulated Annealing*. This method originated in spin glass research [4] and represents a modification of the *Monte-Carlo* optimization methods [5].
- *Neural Networks*. These were originally conceived as models for the dynamics of central nervous systems. Recently, they were rediscovered and are now used in certain optimization problems. The, main field of application, however, is pattern recognition [6].
- *Genetic Algorithms*. These methods mimic biological evolution and apply its principles to optimization problems in other disciplines [7-9].

In this contribution, we shall apply simulated annealing and a genetic algorithm which is derived from the evolution of simple organisms – so-called *prokaryotes*: viruses, bacteria and cyanobacteria – to problems of molecular evolution and compare the results of both techniques.

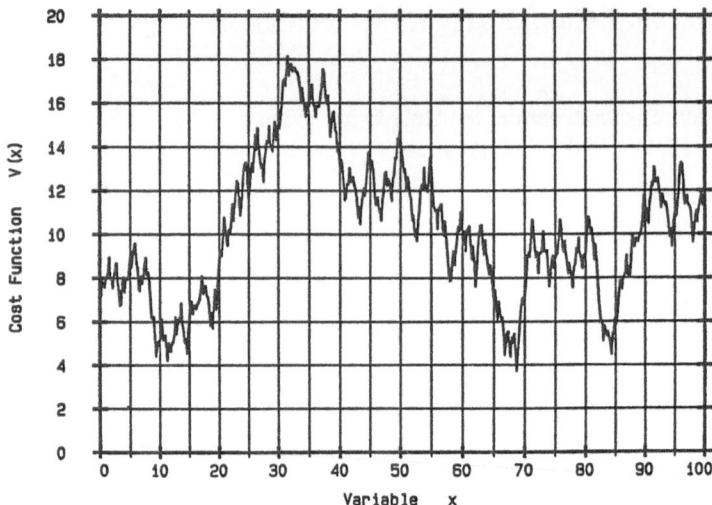

Fig. 1: A cost function $V(\mathbf{x})$ characteristic of a complex optimization problem. An arbitrarily chosen degree of freedom is plotted on the x-axis. The cost function $V(\mathbf{x})$ commonly depends on many variables. It has a great number of maxima. Only the highest peaks, however, are useful as approximate solutions. Typical cost functions have bizarre structures and are often called *rugged* landscapes: neighbouring vectors $\mathbf{x}$ and $\mathbf{x}+\delta\mathbf{x}$ may belong to very different values of the cost function.

Most complex optimization problems share a common feature: one searches for the global maximum – or global minimum – of a highly structured objective function or cost function $V(x_1, x_2, \ldots, x_n)$ of many variables. The number n may be as large as several thousand or more. In some cases the qualitative behaviour in the limit $\lim n \to \infty$ is of interest. Using vector notation, $\mathbf{x} = (x_1, x_2, \ldots, x_n)$, we can define the problem of finding the maximum of the cost function by

$$\mathbf{x}_m : \qquad V(\mathbf{x}_m) \;=\; \max\{V(\mathbf{x})\} \tag{1}$$

The main difficulty of this search is illustrated in Fig. 1. A typical cost function for a complex optimization problem has many maxima widely distributed with respect to · height. Only a small percentage of these peaks is of interest. Most of them lie at small values of the cost function and are useless as approximate solutions to the optimization problem.

An example of complex optimization in physics is provided by the search for a minimum of the spin glass Hamiltonian [10,11]

$$\mathcal{H} \;=\; \sum_{i=1}^{n} h_i s_i + \sum_{i=1}^{n-1}\sum_{j>i}^{n} J_{ij}\, s_i s_j \;. \tag{2}$$

Variables of this cost function are the individual spins s_i which can take two values: $s_i \in \{+1, -1\}$; $i = 1, 2, \ldots, n$. The parameters h_i represent the local magnetic fields experienced by the individual spins s_i in a spin glass. The spin-spin constants J_{ij} are randomly distributed. A state of the spin system – a *configuration* – is described by vector $S_j = (s_1^{(j)}, s_2^{(j)}, \ldots, s_n^{(j)})$. For some applications it is useful to define a *configuration space* whose elements are the 2^n different vectors S_j. This configuration space is identical to the *sequence space* of binary sequences of the same length (n) shown in Fig. 3. Configurations correspond to the corners of a hypercube of dimension n. The edges connect configurations which are transformed into each other by a single spin flip.

The challenge is to find the configuration (S) which minimizes the Hamiltonian. The complexity, is caused in essence, by the coincidence of two conditions:

- exponentially increasing diversity of spin configurations as the number of spins increases and
- lack of translational invariance of the Hamiltonian due to randomness in sign and value of the coupling constants J_{ij} in the spin glass.

In contrast to spin glasses, magnetic *Ising* systems do not meet the second condition. All non-zero coupling constants are set equal. In the one, dimensional nearest neighbour case, these assumptions yield: $J_{ij} = J \cdot \delta_{j,i\pm1}$. We make use here of Kronecker's symbol: $\delta_{ij} = 1$ if $i = j$ and $\delta_{ij} = 0$ if $i \neq j$ otherwise. Then, the cost function described by the Hamiltonian (2) has only a single minimum or two degenerate minima, and the optimization problem is accordingly fairly simple. Unlike in *Ising* systems, it is impossible to achieve an optimum configuration by regular orientation of spins in spin glasses. Randomly distributed coupling constants that vary in sign, cause *frustration* [12] : favourable orientation of one spin relative to a second inevitably creates an unfavourable orientation relative to others. Frustration thus leads to the appearance of many minima of very different depth in the spin glass Hamiltonian.

Optimization by *simulated annealing* mimics the cooling of a spin glass. At high temperatures, all configurations have essentially the same probability of being realized. Slow cooling causes spin systems to approach configurations of low energy. The cooling process is used to modify the Monte-Carlo search for the global maximum of a cost function $V(x)$. The search is initiated by a start vector x_0 and new vectors are created by variation of x_0 with a random element. Depending on the new value of the cost function and the selection principle applied, the new vector is either accepted – x_1 replaces x_0 – or rejected. Repetition of this procedure leads to a sequence of temporarily accepted solutions

$$x_0 \rightarrow x_1 \rightarrow \ldots \rightarrow x_k \rightarrow x_{k+1} \rightarrow \ldots (\rightarrow x_m) \ , \tag{3}$$

which converge eventually to the optimal solution. The probability of acceptance of one particular step, say $x_k \rightarrow x_{k+1}$, is determined by the relation

$$\text{Prob}(x_k \rightarrow x_{k+1}) = \begin{cases} 1 & \text{if } V(x_k) \geq V(x_{k+1}) \\ \exp\left\{ \left(V(x_k) - V(x_{k+1})\right) \big/ T \right\} & \text{if } V(x_k) < V(x_{k+1}). \end{cases}$$

Equation (3) implies that a new solution is always accepted if it is better than, or as good as the current one. Worse solutions can be accepted too, but with a certain probability which decreases exponentially with the quality of the solution. In this manner, the optimization avoids being trapped in low-lying local maxima. The probability of acceptance of a worse solution is given by a Boltzmann weight which uses a formal temperature T. The success of *simulated annealing* depends on the choice of an appropriate temperature program. A good program starts at sufficiently high temperature to reduce the sensitivity to starting conditions and then cooling is performed as close as possible to the formal thermodynamic equilibrium.

Evolutionary adaptation in biology can be understood as a complex optimization problem [13]. The object to be varied is a polynucleotide molecule – DNA or, in some classes of viruses, RNA. It is commonly called the *genotype*. The genotype is not evaluated as such, it is first mapped into a *phenotype* and then the selection principle operates on phenotypes. Charles Darwin's great merit was to find the universal mechanism of phenotype evaluation in nature: *natural selection* implies that those variants which have more – and/or more fertile – descendants in future generations increase in number, whereas their less efficient competitors decrease. Efficiency in the natural evaluation process refers exclusively to reproductive success. The variant which produces the largest number of fertile descendants ultimately replaces all competitors and is selected. In population genetics, the notion of *fitness* was introduced as a quantitative measure for the capability of being selected. The cost function of evolutionary optimization is therefore often characterized as a *fitness landscape.* We shall show, however, that fitness in the sense of Darwin's *survival of the fittest* cannot be assigned to a single genotype. It is a collective property. Therefore we suggest the use of the notion of *value landscape* rather than fitness landscape as an appropriate synonym for cost function in biology.

It is impossible to derive analytic expressions for realistic value landscapes in biology. This holds true even for the most simple conceivable cases like folding and replication of nucleic acid molecules. Here, we study a model of this process, in which secondary structures are computed from known genotypes by means of a folding algorithm.

Genetic algorithms mimic evolutionary optimization. A *population* of test solutions is set up, individual solutions are *replicated* and subject to variations. The better a given solution, the more descendants it has. Two types of variations are commonly considered: mutations – variations within a solution – and recombinations – exchange of parts between different solutions. Selection is achieved by constant population size. Excess production is compensated by an appropriate dilution flux.

2. Genotype and Phenotype

In view of their chemical structure, nucleic acids are best visualized as *strings* or sequences of symbols. There are four classes of symbols: in ribonucleic acid (RNA) they consist of G(guanine), A(adenine), C(cytosine) and U(uracil). Double helical structures (Fig. 2) are particularly stable in themodynamical terms. The symbols are grouped into two base pairs which constitute complementarity relations: $G \equiv C$ and $A = U$. All other combinations of symbols opposing each other in the double helix cause low stability or instability.

In order to illustrate the source of ruggedness in the value landscapes of RNA folding and replication, we distinguish the genotypic sequences or strings I_k and the phenotypic [14] spatial structures $\mathcal{G}_k(I_k)$. The phenotype is formed through folding into a three dimensional structure as shown in Fig. 2. The driving force for this spontaneous process in an aqueous solution of proper ionic strength is the minimization of Gibbs'

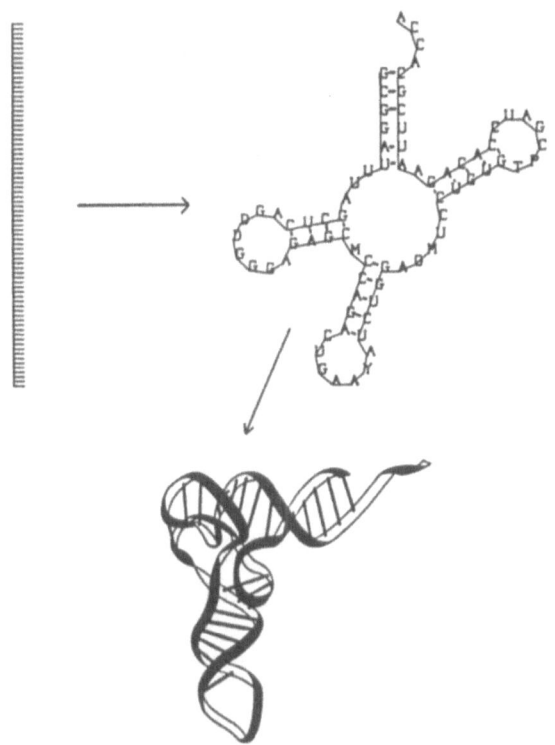

Fig. 2: Folding of a polyribonucleotide chain into the two dimensional secondary and the three dimensional tertiary structure of the RNA molecule. *Phenylalanyl-transfer-RNA*, an RNA with a chain length of $\nu = 76$ bases, is chosen as an example. The symbols **D, Y, M, T** and **P** are used for modified bases which are unable to form base pairs. The major·driving force for folding into secondary structures is the formation of thermodynamically favoured double helical substructures which contain stacks of complementary base pairs: **G≡C** and **A=U**. It is important to note that in double helices the two nucleotide strands run in opposite direction.

free energy through base pairing, stacking and other types of molecular association. Polyribonucleotide folding can be partitioned into two steps:
- folding of the string into a quasiplanar – two dimensional – secondary structure through formation for as many base pairs as possible and
- formation of a spatial – three dimensional structure from the planar folding pattern.

The first of these two steps is modelled much more easily than the second. Efficient algorithms are available for the computation of RNA secondary structures [15]. The prediction of tertiary structures of RNA molecules is much harder. At present there are no reliable theoretical models available for this purpose.

3. Evaluation of Phenotypes

The evaluation criterion of Darwin's principle of selection in stationary populations is the mean number of descendants: variants with more descendants than the mean increase as a percentage of the total population. Less successful competitors decrease in number until they die out. Elimination of less efficient variants causes the mean to increase. Then more and more variants fall under the failure criterion and ultimately only a single fittest type remains as expressed by the common – and often misused – catchphrase of the *survival of the fittest*. In order to make the model quantitative and to develop it further, a measure of *fitness* is required. This measure has to be defined independently of the outcome of selection, in order to avoid the long-known vicious circle of the *survival of the survivor*. In principle, chemical reaction kinetics is in a position to provide independent access to the selection criterion if applied to the biological multiplication process.

The main difficulty in the experimental investigation of Darwin's principle is caused by the enormous complexity of all organisms – even bacteria – which is presently too great to study multiplication using the methods of physics and chemistry. The simplest system in which selection could be observed consists of RNA molecules which are multiplied *in vitro* by a replication essay [16]. This essay contains an RNA replicase, which replicates the RNA of the bacteriophage $Q\beta$ in cells of the bacterium *Escherichia coli*, together with an excess of all molecular materials necessary for the synthesis of RNA. By means of the $Q\beta$ system, the kinetics of RNA replication was explored in great detail [17]. These investigations can be seen as the basis of a molecular model of evolution [18,19] which we now briefly discuss.

Darwin's principle can be formulated on the molecular level by means of kinetic – ordinary – differential equations. The variables are the concentrations of genotypes: $[I_k] = c_k(t)$. The appropriate measure of genotype frequencies in populations are relative concentrations normalized to 1:

$$x_k(t) = c_k(t) \bigg/ \sum_{i=1}^{n} c_i(t) \quad \text{with} \quad \sum_{i=1}^{n} x_i(t) = 1 . \tag{4}$$

Two classes of chemical reactions determine the time dependence of relative concentrations: replication with rate constant A_k

$$(\text{A}) + \text{I}_k \xrightarrow{\quad A_k \quad} 2\,\text{I}_k ,$$

and degradation with rate constant D_k

$$\text{I}_k \xrightarrow{\quad D_k \quad} (\text{B}) ; \quad k = 1, 2, \ldots, n .$$

The symbols **A** and **B** denote low molecular weight materials which are either required for RNA synthesis or which are produced by degradation, respectively. The materials for the synthesis are assumed to be present in excess and their concentrations are essentially constant. Concentrations of degradation products do not enter the kinetic equations, since the process is irreversible under the conditions of the experiment. Hence, neither the concentrations of **A** nor those of **B** represent variables in the kinetic equations – and both symbols were put in parentheses accordingly.

The kinetic differential equations depend only on differences in the rates of synthesis and degradation. It is therefore useful to introce an *excess production*, $E_k = A_k - D_k$, of the individual genotypes I_k:

$$\frac{dx_k}{dt} = (E_k - \bar{E})\, x_k \quad \text{with} \quad \bar{E}(t) = \sum_{i=1}^{n} x_i E_i\,; \quad k = 1,\ldots,n\ . \tag{5}$$

The mean excess production of the entire population is denoted here by $\bar{E}(t)$. It represents the molecular analogue of the mean number of descendants mentioned initially as the evaluation criterion for phenotypes.

The solutions of the differential equation (5) are given by

$$x_k(t) = x_k(0)\,\exp(E_k t) \Big/ \sum_{i=1}^{n} x_i(0)\,\exp(E_i t)\,; \quad k = 1,\ldots,n\ . \tag{6}$$

Provided that we waited for a sufficiently long time, all the variables except one would have approached zero: $\lim_{t\to\infty} x_j(t) = 0$ for all $j \neq m$. The only non-vanishing variable, $\lim_{t\to\infty} x_m(t) = 1$, describes the concentration of that variant which has the maximum excess production, $E_m = \max\{E_i;\ i = 1,\ldots,n\}$, and hence reproduces itself most efficiently. Ultimately, only the fittest genotype I_m remains.

The Darwinian scenario, as well as conventional population genetics is founded on the assumption that mutations are rare events. Selection is fast – occuring on a time scale which is set by the rate constant of replication, or multiplication in general – and evolutionary adaptation is slow, since it is determined by the rate at which advantageous mutants are formed. Only then, can we expect the concept of *survival of the fittest* to reflect reality. Evolutionary optimization is characterized as a sequence of genotypes which were fittest at the instant of their selection.

In the light of the knowledge provided by present day molecular biology two points of the simple evaluation scheme have to be modified:

- The molecular biology of viruses – and to some extent also that of bacteria – has shown that mutations occur much more frequently than the classical scenario assumes. In the case of frequent mutations, no single fittest type survives. A distribution of genotypes generally consisting of a most frequent sequence, the *master sequence*, and its closely related mutants is selected instead. This distribution was called a *quasispecies* [19] in order to point at some analogy to the notion of a species in higher organisms.

- A high precentage of mutants is selectively neutral. Their proliferation in natural populations is the major source of information for phylogenetic studies on the molecular level. The dynamics of neutral selection is a stochastic phenomenon. It can be predicted by probability calculus only. Stochasticity is of general importance when particle numbers are small.

Proliferation of neutral mutants is generally observable in finite populations only. It can be modeled by means of stochastic processes but not by kinetic differential equations. Every model which uses kinetic differential equations is valid in the limit of large – in principle, infinitely large – particle numbers. For most chemical reactions, this is no real restriction, since all relevant results refer to 10^{10} or more molecules. In the domain of molecular evolution, the limit of large numbers is not generally reached. Every new genotype produced by mutation is at first present in a single copy only and the initial phase of growth is also dominated by stochastic phenomena.

The first problem has been extensively dealt with in the theory of the molecular quasispecies [20-22] . We restrict this contribution to a brief account of the most important results. The second question cannot be treated here in sufficient detail and a review of the theory of neutral mutants is found in [23]. The computer simulations introduced in section 5 also comprise the stochastic effects of evolution.

In the frequent mutation scenario, mutant production must be accounted for explicitly. Using Q_{ki} for the frequency at which the genotype I_k is synthesized as an error copy of the sequence I_i, the reactions leading to mutants can be written as

$$(\mathbf{A}) + \mathbf{I}_i \quad \xrightarrow{A_i Q_{ki}} \quad \mathbf{I}_k + \mathbf{I}_i .$$

Adding these reactions to error-free replication and degradation, we end up with the kinetic differential equations of the replication-mutation system:

$$\frac{dx_k}{dt} = \sum_{i=1}^{n} A_i Q_{ki} x_i - \left(D_k + \bar{E}(t) \right) x_k ; \quad k = 1, \ldots, n . \tag{7}$$

We note that the mean excess production $\bar{E}(t)$ is defined precisely as in equation (5). Since every descendant is either a correct copy of the parental genotype or a mutant, a conservation law holds: $\sum_{k=1}^{n} Q_{ki} = 1$. The *mutation matrix* $Q \doteq \{Q_{ki}\}$ is a stochastic matrix. In the case of symmetric mutation frequencies, $Q_{ik} = Q_{ki}$, it is a bistochastic matrix.

Equation (7) can be solved explicitly too [24,25]. As in (6) the solutions are superpositions of exponential functions, but now they are obtained as linear combinations of the eigenvectors $\vec{\ell}_j$ of a *value matrix* $W \doteq \{W_{ij} = A_j \cdot Q_{ij} - D_i \cdot \delta_{ij}\}$. The coefficients are time dependent: $u_j(t)$. The solution curves have the same general form as those shown in equation (6): the variables $x_k(t)$ have to be replaced by the coefficients $u_j(t)$ and and the eigenvectors of the value matrix, λ_j, appear in the exponential functions instead of the excess productions E_k.

The quasispecies Γ_0 can now be defined in precise mathematical terms. It is the combination of genotypes which is described by the dominant eigenvector of the value matrix, $\vec{\ell}_0 = (\ell_{10}, \ell_{20}, \ldots, \ell_{n0})$ – this is the eigenvector which belongs to the largest eigenvalue: $\lambda_0 > \lambda_1 \geq \lambda_2 \geq \ldots \geq \lambda_n$. The quasispecies can be written as

$$\Gamma_0 = \ell_{10} \mathbf{I}_1 \oplus \ell_{20} \mathbf{I}_2 \oplus \ldots \oplus \ell_{n0} \mathbf{I}_n . \tag{8}$$

The components ℓ_{k0} measure the contributions of the individual genotypes I_k to the quasispecies. Their size is determined to a first approximation by the frequencies of mutation from the master sequence I_m to the genotype I_k and by the difference in selective values:

$$\ell_{k0} = \frac{Q_{km}}{W_{mm} - W_{kk}}; \quad k = 1, 2, \ldots, n; \ k \neq m; \ W_{kk} < W_{mm} . \tag{9}$$

Here, we characterize the diagonal elements of the value matrix, $W_{kk} = A_k Q_{kk} - D_k$, as selective values. The master sequence is usually the genotype with the largest selective value – for exceptions see [26] and Section 4.

Making use of the structure of matrix W, it can be shown that all components of the eigenvectors $\vec{\ell}_0$ are positive. The sum of relative concentrations is constant, $\sum_{k=1}^{n} x_k(t) = 1$, and the same holds true for the sum of the coefficients of the eigenvectors, $\sum_{j=1}^{n} u_j(t)$. Accordingly, the contributions of all eigenvectors except $u_0(t)$, that of the dominant eigenvector $\vec{\ell}_0$ vanish in the limit $t \to \infty$. After a sufficiently long time, the mutant distribution converges towards the quasispecies.

The extension of the conventional selection model to the frequent mutation scenario has two important consequences:

- The goal of the selection process is not the fittest genotype but a clan of genotypes consisting of a master sequence – eventually of two or more master sequences – together with its frequent mutants. *Fitness* in the sense of Darwin's selection theory is a property of this clan, called the *quasispecies*. Fitness is not a property of the master sequence alone.

- Quasispecies in real – finite – populations become unstable at large mutation frequencies. Too many errors accumulate during replication and cause the loss of the master sequence and its mutants. Then, there exists no stable stationary mutant distribution.

The latter phenomenon is best illustrated in an abstract vector space of dimension ν – here ν is the length of the sequences. Those vectors which correspond to individual sequences are elements of the so-called *sequence space* [27]. Examples of low-dimensional sequence spaces of binary sequences are shown in Fig. 3. Sequence spaces are identical to the configuration spaces of spin systems. In the case of a sufficiently accurate replication, the support of the population (the set of all points whose sequences are present in the population) is stationary. The mutant distribution is localized [28]. When the error frequency exceeds some critical value, the population starts to walk randomly through sequence space. The surprising result of the theory is not the phenomenon as such, it is the sharpness of the transition from a localized *quasispecies* to a drifting mutant distribution which has much in common with phase transitions. It is quite justifiable to speak of a critical error threshold at which the population becomes delocalized.

The relationship between the error threshold of replication and phase transitions is much deeper than a mere similarity. It was shown that statistical mechanics of equilibrium phase transitions in spin systems is equivalent to the replication-mutation dynamics of binary sequences [29,30] . The localization-delocalization threshold of populations in sequence space corresponds to the order-disorder transition transition in spin systems. Examples are transitions from ferromagnetic or antiferromagnetic phases into the paramagnetic state.

Localization of quasispecies in sequence space is of fundamental importance for the evaluation of genotypes. Only localized populations conserve the information stored in sequences over sufficiently many generations. Only they can be evaluated by natural selection and are suitable for evolution. Delocalized, drifting populations cannot adapt evolutionarily to the environmental conditions.

4. A Simple Model

Despite the enormous progress in the physical and chemical characterization of selection and evolutionary adaptation – which was achieved by means of the $Q\beta$ RNA replication essay – many questions cannot be addressed experimentally as yet. Amongst other problems, there is the experimental determination of the distribution of selective values in sequence space and the analysis of mutant distributions – during adaptation and in

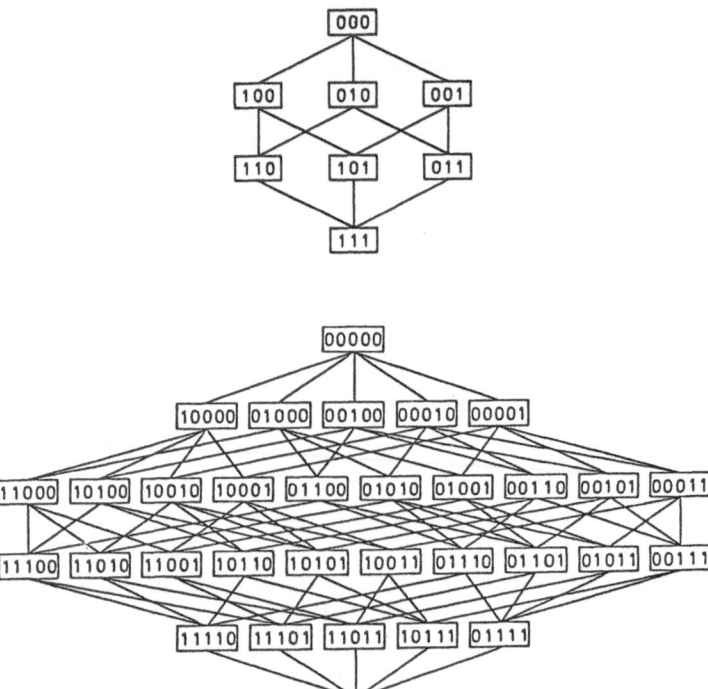

Fig. 3: Sequence spaces of binary sequences of chain lengths $\nu = 3$ and $\nu = 5$. All pairs of sequences with Hamming distance $d = 1$ are connected by a straight line. The object drawn in this manner is a *hypercube* of dimension ν. The Hamming distance is the number of digits by which two sequences differ. Such a *vector-(-point-)-space* was used first by Richard Hamming [27] in information theory. The sequence space is identical with the configuration space of a spin system of the same dimension. The edges of the hypercube connect configurations which can be transformed into each other by a single spin flip.

the case of stationarity – which by far exceed present day capacities. Answers to these questions are, however, fundamental to an understanding of the evolutionary process. A simple model system was conceived suitable for numerical calculations and computer simulation [31,32]. In essence, the model is based on three assumptions:

- Binary $(0,1)$ sequences are considered instead of real polynucleotides. As with RNA, non-neighbouring elements of the polymer chain interact mainly through complementary base pairing, $0 \equiv 1$, and the stability of secondary structures is essentially determined by the number of base pairs.
- Variation of sequences is restricted to point mutations which exchange the two symbols at a given position in the polymer: $0 \to 1$ or $1 \to 0$. The chain length thereby remaining constant.
- The accuracy of polymer synthesis – measured in terms of the frequency at which the correct symbol is incorporated into the growing polymer chain, $1 \geq q \geq 0$ – is assumed to be independent of the position in the chain.

As a consequence of these three assumptions, we can express the frequencies of all mutations $I_i \rightarrow I_k$ – which are the previously mentioned elements of the mutation matrix Q – by the chain length ν, the Hamming distance of the two sequences, $d(I_k, I_i) = d_{ki}$ and a single parameter for the accuracy of replication, q:

$$Q_{ki} \;=\; q^\nu \left(\frac{1-q}{q}\right)^{d_{ki}} . \tag{10}$$

The accuracy parameter of replication, q, is a measure of the precision with which individual digits can be incorporated into the polymer chain: $q = 1$ implies error-free copying – every symbol in the copy is identical to that in the template. In the case $q = 0$, it is always the opposite symbol that is incorporated – we are dealing with complementary copying which reminds one of the conventional photographic technique. Instead of positive and negative, *plus-* and *minus*-strands alternate here. Sequences occur in pairs, $(I_k^\oplus, I_k^\ominus)$, in complementary replication. Another extreme case is observed at $q = 0.5$: both symbols, 0 and 1, are incorporated with equal probabilty – and independently of the symbol which is at the corresponding position in the template. The error frequency is at maximum and there is no correlation between the sequences of template and copy. We may characterize this case appropriately as *random replication*.

Most simple distributions of kinetic parameters in sequence space are particularly well suited to illustrate some fundamental properties of quasispecies. We restrict selection to the synthesis of macromolecules and set all degradation rate constants equal: $D_1 = D_2 = \ldots = D_n = D$. The solutions of equations (5) and (7) are invariant to additive constants and we choose $D = 0$ without losing generality. Here we consider two examples of value landscapes for which the structures of quasispecies were studied as functions of the accuracy parameter q [26,31].

4.1. The Single-Peak-Landscape

On the *single-peak-landscape* a higher value of the replication parameter is assigned to the sequence I_0, all other replication parameters are assumed to be equal: $A_0 = \alpha$ and $A_1 = A_2 = \ldots = A_{2^\nu - 1} = 1$. Consequently I_0 is the master sequence $(m = 0)$. The domain $1 \geq q \geq 0$ is partitioned into three ranges where direct, random and complementary replication dominate. The individual zones are separated by narrow transitions – even at chain lengths as short as $\nu = 10$. We may define two error thresholds, q_{min} for direct and q_{max} for complementary replication.

Direct replication dominates in the range $1 \geq q \geq q_{min}$. In the limit $q \rightarrow 1$ the quasispecies converges towards the pure master sequence I_0.

In the range $0 \leq q \leq q_{max}$ complementary replication dominates. At $q = 0$ a master pair consisting of I_0 $(= I_0^\oplus)$ and its complementary sequence $I_0^\ominus$ is selected. For accuracy, parameters $q > 0$, the master pair is accompanied by a distribution of its mutant pairs: $(I_1^\oplus, I_1^\ominus)$, $(I_2^\oplus, I_2^\ominus)$ etc. The goal of selection is a quasispecies of complementary pairs.

Chain elongation – increase in the chain length ν – sharpens the two transitions remarkably. This can be understood as a strong hint that we are dealing with a co-operative process related to phase transitions. In Fig. 4 we show the stationary mutant distribution for binary sequences of chain length $\nu = 10$ in the entire domain $1 \geq q \geq 0$. The stationary concentrations of the individual genotypes are essentially

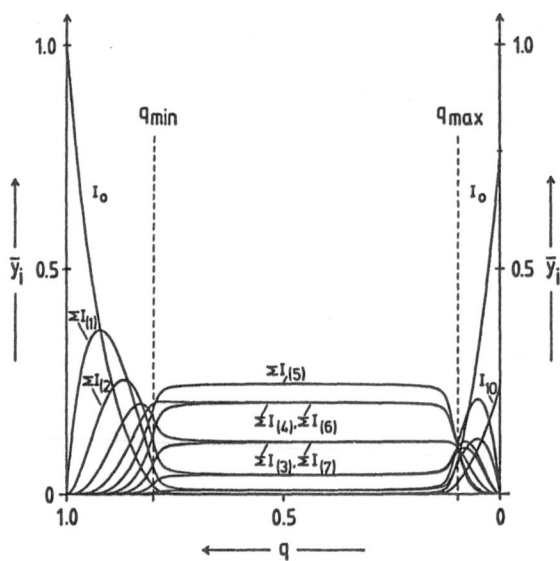

Fig. 4: The stationary mutant distribution called *quasispecies* of binary sequences of chain length $\nu = 10$ on the *single-peak-landscape* as a function of the replication accuracy q. A replication parameter $\alpha = 10$ was used. We plot the relative stationary concentrations of error classes as functions of q. Error classes are the master sequence I_0, all one-error mutants $I_{(1)}$, all two-error mutants $I_{(2)}$, ... , and finally the ν-error mutant $I_{(\nu)}$ of the master sequence. Two transitions occuring at critical values of the replication accuracy, q_{min} and q_{max}, are easily recognized. In the domain of random replication, which lies between the two transitions, all individual genotypes are formed with essentially the same frequency – corresponding to a uniform distribution of sequences. Since we plot concentrations of error classes, the relative frequencies are proportional to the binomial coefficients in this range: $[I_{(k)}] = y_k = \binom{\nu}{k}$.

constant betweem the two critical values, i.e. in the range $q_{min} > q > q_{max}$. In addition, all sequences are present at the same frequency. This implies, however, that stationary solutions cannot be approached by realistic populations when the chain lengths exceed $\nu = 80$. No population can be larger than one mole individuals, and $6 \cdot 10^{23}$ is roughly equal to 2^{80}. When the population is too small, it cannot be localized in sequence space but instead drifts randomly and all genotypes have only a finite lifetime.

4.2. The Double-Peak-Landscape

Two genotypes are characterized by high replication efficiency on *double-peak-land-scapes*. In this case, two sequences are close relatives – having a Hamming distance $d = 1, 2, \ldots$, up to some limit determined by population size, sequence length and structure of the value landscape – they are readily produced as mutants of each other and then a common quasispecies is formed at non-zero error rates, $q < 1$. It consists of the two highly efficient genotypes and a cloud of mutants surrounding both.

The case of more distant peaks appears more interesting. Then, two distinct quasispecies – two distributions without common mutants – can exist. By an appropriate choice of the kinetic parameters, both quasispecies can be stable. Each quasispecies has its own range of stability on the q-axis. These ranges do not overlap; instead they are separated by sharp transitions at critical replication accuracies q_{tr} which again resemble phase transitions.

A *double-peak-landscape*, which is well suited for computation and illustration of transitions between quasispecies, puts the two highest selective values at maximal Hamming distance $d = \nu$ – the two master sequences are complementary. Four different replication parameters are used for the two master sequences and their one-error mutants, in particular

(1) $A_0 = \alpha_0$, for the master seqeunce I_0,
(2) $A_{(1)} = \alpha_1$, for all one-error mutants $I_{(1)}$ – we consider all one-error mutants together as an error class and express this by the subscript "(1)" –
(3) $A_{(\nu)} = \alpha_\nu$ for the second master sequence $I_{(\nu)}$ and
(4) $A_{(\nu-1)} = \alpha_{\nu-1}$ for all one-error mutants $I_{(\nu-1)}$ of $I_{(\nu)}$ – these are the $(\nu-1)$-error mutants of I_0.

The replication constant $A_k = 1$ is assigned to all other sequences.

An illustrative example of quasispecies rearrangement on a *double-peak-landscape* is shown in Fig. 5. The two genotypes with high replication rates were assumed to be almost selectively neutral – the difference in replication rates is very small. The less efficient genotype has more efficient mutants than its competitor. Then, the more efficient master sequence dominates at low, the less efficient at high error rates.

The mechanism of the rearrangement of stationary mutant distributions at some critical replication accuracy, q_{tr}, is easily interpreted. The genotype with the somewhat higher replication parameter and the less efficient neighbours in sequence space dominates at small error rates because the mutation backflow from mutants to master sequence is negligible. With increasing error rates – decreasing q-values – mutation backflow becomes more and more important, and at the critical replication accuracy, q_{tr}, the difference in replication rates of the two master sequences is just compensated by the difference in mutation backflow, further increase in error rates makes the less efficient master sequence superior to that with the higher replication rate constant. The conventional selection principle – as visualized by the rare mutation scenario – has to be modified in one area of primary importance:

- Fitness is a collective property and cannot be attributed to a single genotype. Also, fitness is determined by all sequences of the quasispecies and precisely to the extent in which they contribute to the stationary distribution.

Accordingly, fitness is also an implicit function of the replication accuracy q.

Studies on simple model landscapes are well-suited to the illustration of phenomena like error thresholds or rearrangements of quasispecies. They suffer, of course, from the fact that they have little in common with real value landscapes. In order to be able to proceed towards an understanding of selection and adaptation on the molecular level, we must inevitably investigate more realistic models.

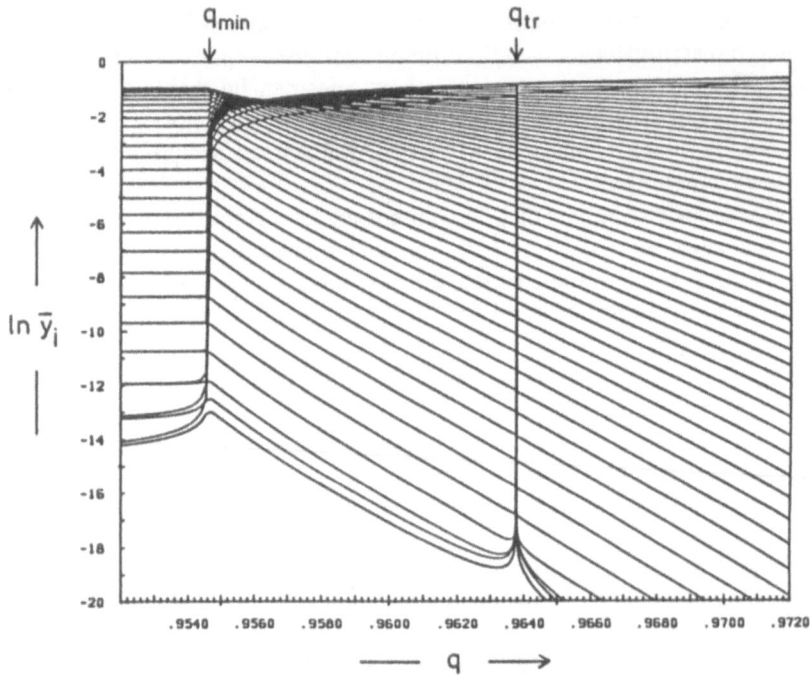

Fig. 5: The stationary mutant distribution called the *quasispecies* of binary sequences of chain length $\nu = 50$ on the *double-peak-landscape* as a function of the replication accuracy q. Relative concentrations of the individual error-classes are plotted on a logarithmic scale in order to illustrate the sharpness of the two transitions. The two almost neutral master sequences are positioned at maximal Hamming distance, $d = 50$. The master sequence with the higher replication efficiency, I_0 with $\alpha_0 = 10$, is surrounded by the less efficient one-error mutants, $\alpha_1 = 1$, whereas the less efficient master sequence, $I_{(50)}$ with $\alpha_{50} = 9.9$, has better one-error mutants, $\alpha_{49} = 2$. Two sharp transitions are observed: a rearrangement of quasispecies at $q_{tr} = 0.9638$, where the two master sequences I_0 and $I_{(50)}$ exchange their roles, and the error threshold at $q_{min} = 0.9546$. I_0 is the most frequent genotype in the range $1 \geq q > q_{tr}$ and $I_{(50)}$ dominates in the neighbouring range $q_{tr} > q > q_{min}$.

5. A Model of a Realistic Value Landscape

Relations between genotypes and phenotypes are highly complex even in·the most simple systems which are performed in test-tube RNA replication and evolution experiments. The phenotype is formed in a many-step process through decoding of the genotype. The outcome of this process is usually unequivocal in a constant environment, but many genotypes may yield the same phenotype.

The genotype-phenotype relation in the model [32,33] considered here is reduced to the folding of a binary string into a secondary structure following a minimum free energy criterion as described in Section 3. In the actual computations, we applied base pairing and base pair stacking parameters of pure G,C-sequences [34,35]. The folding process is restricted to the formation of unknotted planar structures. This restriction is common in all currently used folding algorithms, and is also well justified on an empirical basis of structural data known at present.

An important common feature of all correlations between genotypes and phenotypic properties is the *bizarre* structure of value landscapes. This means that closely related genotypes may – but need not – lead to phenotypes of very different properties. Distant genotypes, on the other hand, may yield almost identical phenotypic properties. This peculiar feature of genotype-phenotype relations is illustrated by means of secondary structures of binary sequences in Fig. 6.

0110001101001001101000001011010011000011101100101010001110010101000100

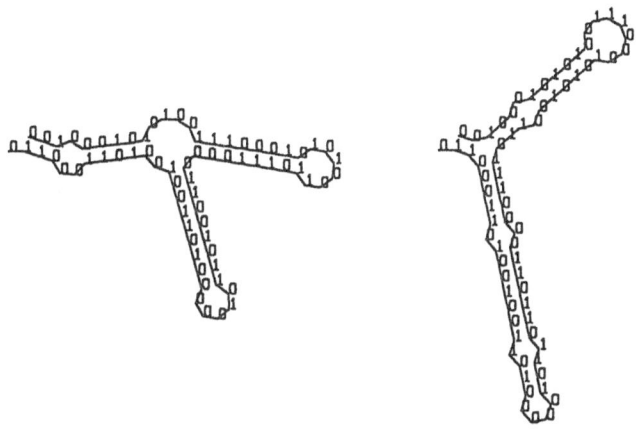

0010110100011010100110111011001100000010100000000001010000101100001101

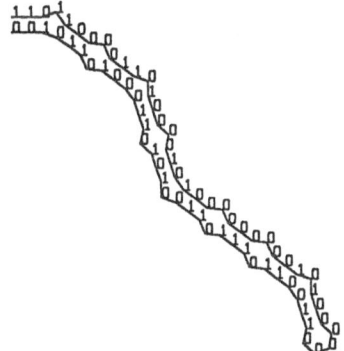

Fig. 6: Examples of secondary structures of binary sequences. One main reason for the *ruggedness* of value landscapes is to be seen in the complexity of the folding process which, presumably, cannot be cast into any simpler algorithm. In the upper part of the figure, we show two sequences of Hamming distance $d=1$ which have very different secondary structures and below that, two rather distant sequences with $d=9$ which fold into indentical secondary structures. Mutated positions are indicated by $*$.

Rate constants of replication (A_k) and degradation (D_k) are particularly important properties in molecular evolution. Existing knowledge of the dependence of these constants on RNA structures is not sufficiently detailed to allow reliable predictions. Nevertheless, it is possible to conceive models which reproduce qualitative dependences correctly. Such an ansatz was used in the computation of a value landscape [32] which describes the distribution of excess productions $(E_k = A_k - D_k)$ in sequence space. In order to approach an optimal excess production the replication rate should be as large as possible $(A_k \rightarrow \max)$ and, at the same time, the degradation rate should be at minimum $(D_k \rightarrow \min)$. The two constants depend on secondary structures in such a way that both conditions cannot be fulfilled: the two trends contradict each other. An increase of A_k is commonly accompanied by an increase rather than a decrease in D_k. We thus observe conflicting structural dependences are remniscent which remind of the frustration phenomenon in spin glasses [12] and we expect a value landscape of high complexity.

In order to explore the details of the value landscape, two kinds of *computer experiments* were carried out:

- A population of 3000 binary sequences of chain length $\nu = 70$ was put on the value landscape in order to study the optimization by means of a genetic algorithm. The accuracy parameter was kept constant: $q = 0.999$. The stochastic dynamics of a chemical reaction network consisting of replication, mutation and degradation processes was simulated by means of an efficient computer algorithm [36]. Sequences produced in excess are removed by a dilution flux.
- The distribution of optimal and near optimal regions of excess production (E_k) in sequence space was investigated by simulated annealing [4]. In this search one- and two-error mutants were admitted as variations of the current strings.

The mean excess production $\bar{E}(t)$ increases steadily – apart from small random fluctuations – during an optimization run with the genetic algorithm and finally approaches a plateau value. The secondary structures obtained in three independent optimization runs which were performed with the genetic algorithm are shown in Fig. 7. Despite identical initial conditions – 3000 all-0-sequences of chain length $\nu = 70$ – the three runs reach different regions in sequence space. These three regions of high excess production are far away from each other. The Hamming distances between these finally selected genotypes are roughly equal, and lie around $d = 30$. Although the kinetic properties of the three optimization products are very similar, they differ largely in sequence and secondary structure.

Depending on population size, replication accuracy and the structure of the value landscape, two different scenarios of optimization were found:

- The population resides for rather a long time in some part of sequence space and moves quickly and in a jumplike manner into another area. The optimal solution is approached by stepwise improvements.
- The population approaches the optimum gradually through steady improvement.

The first scenario is characteristic for small populations, low error rates and distant local maxima in sequence space, whereas optimization dynamics of the second kind dominates in large populations, at high error rates and with nearby local maxima. There is a steady transition from one scenario to the other: the gradual approach

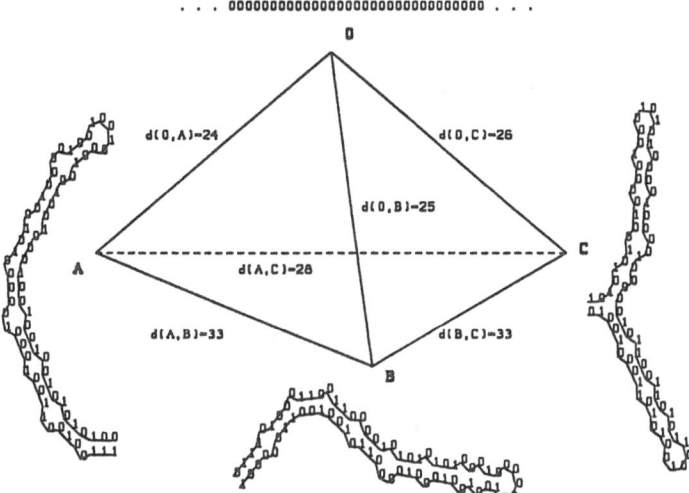

Fig. 7: Secondary structures obtained in three independent optimization runs performed with the genetic algorithm. All three runs started with identical initial conditions, 3000 all-0-sequences of chain length $\nu = 70$, but used different seeds for the random number generator. We show the most frequent genotypes at the end of the optimization processes. Interestingly, these three sequences and the all-0-sequence at the start have pairwise almost the same Hamming distances. The four sequences span an approximate tetrahedron in a three dimensional subspace of the sequence space.

changes, for example, into a sequence of small jumps when the error rate is reduced. Larger jumps appear with a further decrease in the mutation frequency and finally the population is *caught* in some local maximum.

Optimization runs with the genetic algorithm were also carried out at replication accuracies which lie below the critical accuracy for a population of 3000 molecules – the error threshold was found to depend on population size [37] and occurs at higher replication accuracies in smaller populations. No optimization on the value landscape was observed. The populations drift randomly in sequence space. This general behaviour was observed even when we started with *near optimal* genotypes which had been optimized in previous runs. The initial master sequence is first surrounded by a growing cloud of mutants and then is displaced by its own error copies. All other sequences in the population are likewise lost after a sufficiently long time. In other words, every genotype has only a finite lifetime. Therefore, the predictions of the molecular theory of evolution as introduced in Section 3 are also valid in populations as small as a few thousand molecules.

In order to complement optimization by genetic algorithms, the value landscape was investigated by means of simulated annealing [4]. The main aim of this study was to explore the distribution of high maxima of the excess production (E_k) and to compare the model value landscape of polynucleotide replication with a spin glass landscape. The highest possible value of the excess production of binary sequences with chain length $\nu = 70$ can be computed from the criteria used in the evaluation of replication and degradation rate constants: it amounts to $E_{max} = 2245\,[t^{-1}]$ in arbi-

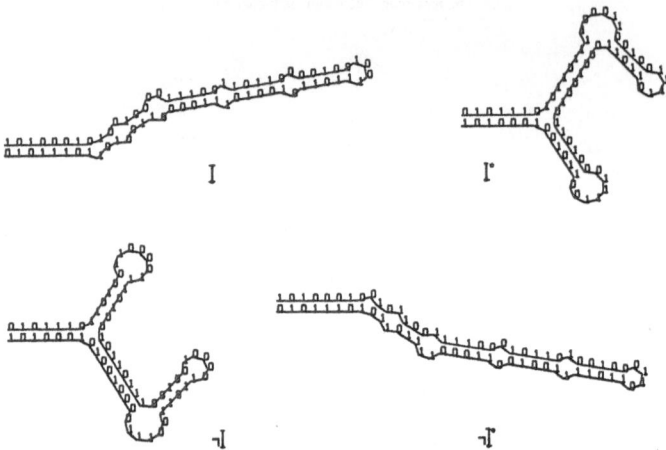

Fig. 8: Symmetry of sequences with respect to secondary sturcture. Every sequence I at distance d from the all-0-sequence origin has a structure which is identical to that of the complementary and inverted sequence $(\neg I^\ominus)$ at distance $\nu - d$ from the origin.

trary reciprocal time units. Such a *best* structure, however, does not occur as a stable folding pattern. All sequences which, in principle, could yield the *best* structure fold into other secondary structures because of the minimum free energy criterion – which of course need not meet a maximum excess production criterion. The highest value of excess production obtained by simulated annealing was $E_{opt} = 2045 \, [t^{-1}]$. This optimal value is degenerate: it is found with ten distinct sequences. The Hamming distances between the ten sequences and the all-0-sequence are given in table 1. The ten genotypes are related in pairs by symmetry: every sequence is transformed into one with identical structure and properties by complementation and inversion – inserting complementary symbols at all positions and swapping both ends (Fig. 8). The remaining five sequences $(I_1, I_3, I_5, I_7$ and $I_9)$ form two pairs, (I_1, I_3) and (I_7, I_9), which are close relatives at Hamming distance $d = 2$ and one *solitary* sequence (I_5). These potential centers for quasispecies are well separated in sequence space – all Hamming distances are larger than $d \geq 30$.

In total, 879 sequences with excess productions $E_k \geq 2011 \, [t^{-1}]$ were identified. Their distribution in sequence space is shown in Fig. 9 by means of a *minimum spanning tree* [38]. The clusters of peaks with high excess production have rich internal structures. High peaks are more probable in the region of other high peaks. Like in mountainous areas on the surface of the earth, we find ridges as well as saddles and valleys separating zones of high excess production.

The distribution of near optimal configurations of the spin glass Hamiltonian (2) shows characteristic features of *ultrametricity*. Arbitrarily chosen triangles of three near optimal configurations are either equilateral or isosceles with small bases. The distribution in sequence space of the high clusters of excess production, on the other hand, shows no detectable bias towards ultrametricity. This distribution of clusters does not deviate significantly from a random distribution. Two factors may be responsible for the difference between spin glass Hamiltonians and value landscapes of polynucleotides. Either binary sequences with a chain length $\nu = 70$ are too short to

Table 1: The Hamming distances between the ten *best* sequences ($I_1,\ldots,I_{10}$) obtained by simulated annealing on the value landscape of the excess production (E_k) of binary sequences and their distance from the all-**0**-sequence (I_0)[t]).

| | Binary Sequences | | | | | | | | | |
	I_1	I_2	I_3	I_4	I_5	I_6	I_7	I_8	I_9	I_{10}
I_0	30	40	30	40	37	33	36	34	36	34
I_1		30	2	32	43	45	34	40	32	38
I_2			32	2	45	43	40	34	38	32
I_3				30	41	43	32	38	30	36
I_4					43	41	38	32	36	30
I_5						24	29	35	31	37
I_6							35	28	37	31
I_7								30	2	32
I_8									32	2
I_9										30

[t] The binary sequences are arranged in pairs of complementary and inverted sequeneces: I_2 is the inverted complementary sequence of I_1 ($I_2 = \neg I_1^{\ominus}$), I_4 is inverse complementary to I_3 ($I_4 = \neg I_3^{\ominus}$), etc.

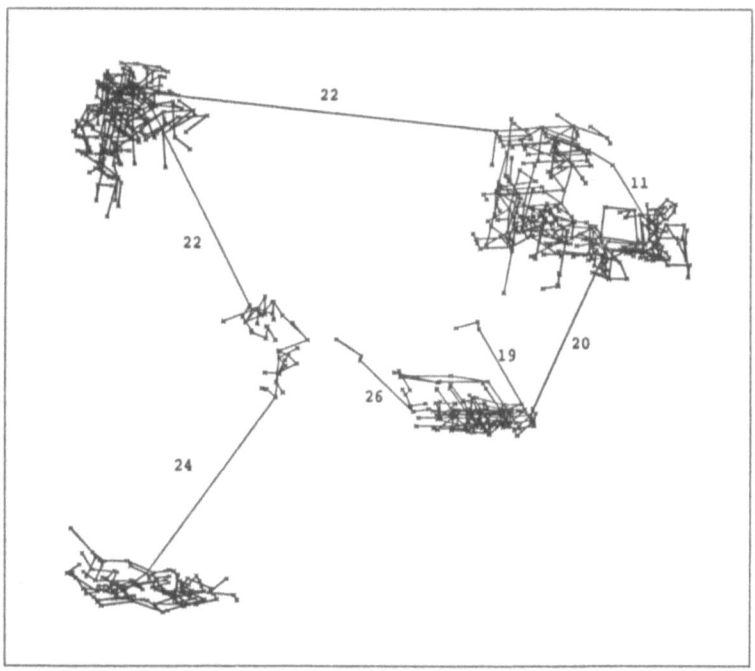

Fig. 9: Clustering of near optimal values of the excess production (E_k). The projection shows the *minimum spanning tree* of the frozen out optima of seven annealing runs operating under identical conditions (except random number seeds). The optimal regions reached by each run consist of an ensemble of selectively neutral mutants and qualify as clusters in the minimum spanning tree. The shortest Hamming distances between clusters are indicated.

reveal higher order structures in the value landscape, or polynucleotide folding and evaluation of folded structures have some fundamental internal structure which is different from that intrinsic to the spin glass Hamiltonian.

Finally, it is worth mentioning that optimization of excess production (E_k) by simulated annealing was not only faster – measured in CPU computer time units – but also led to significantly higher values than the genetic algorithm. This may be, at least in part, a result of the constant environmental conditions applied here. A constant enviroment favours algorithms with little or no memory. Storing of previous *best* solutions – as is naturally done by populations – is advantageous only in variable environments.

6. Concluding Remarks

Rugged value landscapes derived from RNA secondary structures – modeled here by means of binary sequences – have much in common with energy distributions in spin glasses. It seems that (some) complex optimization problems stemming from different scientific disciplines have a common formal origin. Eventually, future developments will lead to a notion of *complexity* which is universal in nature. Apart from the apparent similarities there are, of course, substantial differences in the microscopic origins of complexity. In spin glasses, spin-spin coupling constants are randomly distributed – or they are at least much less regular than in solids – and this leads to *frustration*. In biological evolution, complexity can be traced back to the properties of polynucleotides. In the simplest cases, it is the folding of strings into graphs which follows the rules of molecular biophysics – base pairing and stacking of base pairs such that they yield a minimal free energy – and which creates the complex structure of value landscapes. The notion of frustration can be used here too: structural restrictions have to be fulfilled and thus it is not usually possible to form the maximal number of base pairs.

Frustration appears again when folded structures are evaluated according to a maximum excess production criterion $(E_k = A_k - D_k)$. Here, the origin of frustration is twofold. Firstly, according to model assumptions – certainly endorsed by reality – the genotypes which replicate fastest are also readily degraded. Genotypes which are particularly stable do not replicate very well. Optimization of excess production implies a search for a non-trivial compromise between these two trends.

The second source of frustration is typical of biological problems and was found, – surprisingly – even in our simple model. The *best* structure, having maximal excess production, cannot be realized due to principal restrictions. All secondary structures have to meet a minimum free energy criterion, and no genotype was found which can fulfil both criteria simultaneously; maximum excess production and minimum free energy. Boundary conditions restrict the wealth of possible solutions.

An important question concerns the significance of results derived from binary sequences for realistic RNA molecules representing (G,A,C,U)-sequences. Depending on the thermodynamic parameters used, binary sequences either correspond to pure (G,C)-sequences or pure (A,U)-sequences. In addition, RNA molecules can be described by binary sequences twice as long. Each base is then encoded by two binary digits, for example by $G=11$, $A=10$, $U=01$ and $C=00$. Complementarity relations hold – with the restriction that odd numbered positions have to be opposite to odd numbered positions in double helical structures, and the same holds true for the even numbered positions – but the evaluation of Hamming distances has to be modified and is not as straight forward as with binary sequences. The most important difference, however, concerns the richness in structures, which is certainly greater with binary sequences. Point mutations in true RNA molecules will therefore be somewhat less likely to lead to drastic changes in secondary structures than they do for binary sequences. Nevertheless we can expect all qualitative findings reported here to remain valid in four base sequences.

The occurrence of rugged value landscapes was found to be a fundamental property of biological macromolecules which can be traced back to the process of folding strings of digits into three dimensional molecular structures. This fact is the ultimately responsible for the extreme complexity of the transfrormation process from genotype to phenotype – even in the most simple cases. Thus unfolding of genotypes cannot be reduced to simple algorithms. Genotype-phenotype relations are an important – if not the most important – source of complexity in biology.

Acknowledgements

The work reported here was supported financially by the *Fonds zur Förderung der wissenschaftlichen Forschung in Österreich* (Projects No.5286 and No.6864), the *Stiftung Volkswagenwerk* (B.R.D.) and the *Hochschuljubiläumsstiftung Wien*. Numerical computations were performed on the IBM 3090 mainframe of the *EDV-Zentrum der Universität Wien* as part of the IBM Supercomputing Program for Europe. The Assistance of Dr.Walter Fontana and Mag.Wolfgang Schnabl in preparing the figures and reading the manuscript is gratefully acknowledged.

References

1. J.L.Gersting. *Mathematical Structures for Computer Science.* 2nd Ed., pp.501-504. W.H.Freeman & Co., New York 1987.

2. M.R.Garey, D.S.Johnson. *Computers and Intractability: A Guide to the Theory of NP-Completeness.* W.H.Freeman & Co., San Francisco 1979.

3. D.G.Bounds. *Nature* 329(1987), 215-219.

4. S.Kirkpatrick, D.C.Gelatt, M.P.Vecchi. *Science* 220(1983), 671-680.

5. N.Metropolis, A.Rosenbluth, M.Rosenbluth, A.Teller, E.Teller. *J.Chem.Phys.* 21(1952), 1087-1092

6. J.J.Hopfield, D.W.Tank. *Biol.Cybern.* 52(1985), 141-152.

7. I.Rechenberg. *Evolutionsstrategie.* Friedrich Frommann Verlag, Stuttgart und Bad Cannstatt 1973.

8. J.H.Holland. *Adaptation in Natural and Artificial Systems.* The University of Michigan Press, Ann Arbor 1975.

9. D.E.Goldberg. *Genetic Algorithms in Search, Optimization and Machine Learning.* Addison-Wesley Publ.Co., Reading (Mass.), 1989.

10. S.Edwards, P.W.Ànderson. *J.Phys.F* **5**(1975), 965-974.

11. K.Binder, A.P.Young. *Rev.Mod.Phys.* **58**(1986), 801-976.

12. P.W.Anderson. *J.Less-Common Metals* **62**(1978), 291-294.

13. S.Kauffman, S.Levin. *J.theor.Biol.* **128**(1987), 11-45.

14. Here we generalize the notion of phenotype commonly used for organisms to molecules.

15. M.Zuker, P.Stiegler. *Nucleic Acids Res.* **9**(1981), 133-148.

16. S.Spiegelman. *Quart.Rev.Biol.* **4**(1971), 213-253.

17. C.K.Biebricher, M.Eigen. *Kinetics of RNA Replication by Qβ Replicase.* In: E.Domingo, J.J.Holland, P.Ahlquist, eds. RNA Genetics. Vol.I: RNA-directed Virus Replication, pp.1-21. CRC Press. Boca Raton (Florida), 1988.

18. M.Eigen. *Naturwissenschaften* **59**(1971), 465-523.

19. M.Eigen, P.Schuster. *Naturwissenschaften* **64**(1977), 541-565.

20. M.Eigen, J.McCaskill, P.Schuster. *J.Phys.Chem.* **92**(1988), 6881-6891.

21. M.Eigen, C.K.Biebricher. *Sequence Space and Quasispecies Distribution.* In: E.Domingo, J.J.Holland, P.Ahlquist, eds. RNA Genetics. Vol.III: Variability of Virus Genomes, pp.212-245. CRC Press. Boca Raton (Florida), 1988.

22. M.Eigen, J.McCaskill, P.Schuster. *Adv.Chem.Phys.* **75**(1989), 149-263.

23. M.Kimura. *The Neutral Theory of Molecular Evolution.* Cambridge University Press. Cambridge (U.K.), 1983.

24. C.J.Thompson, J.L.McBride. *Math.Biosc.* **21**(1974), 127-142.

25. B.L.Jones, R.H.Enns, S.S.Ragnekar. *Bull.Math.Biol.* **38**(1976), 12-28.

26. P.Schuster, J.Swetina. *Bull.Math.Biol.* **50**(1988), 635-660.

27. R.W.Hamming. *Coding and Information Theory.* 2nd Ed., pp.44-47. Prentice Hall. Englewood Cliffs (N.J.), 1986.

28. J.S.McCaskill. *J.Chem.Phys.* **80**(1984), 5194-5202.

29. L.Demetrius. *J.Chem.Phys.* **87**(1987), 6939-6946.

30. I.Leuthäusser. *J.Statist.Phys.* **48**(1987), 343-360.

31. J.Swetina, P.Schuster. *Biophys.Chem.* **16**(1982), 329-353.

32. W.Fontana, P.Schuster. *Biophys.Chem.* **26**(1987), 123-147.

33. W.Fontana, W.Schnabl, P.Schuster. *Phys.Rev.A*, submitted.

34. M.Zuker, D.Sankoff. *Bull.Math.Biol.* **46**(1984), 591-621.

35. S.M.Freier, R.Kierzek, J.A.Jaeger, N.Sugimoto, M.H.Caruthers, T.Neilson, D.H.Turner. *Proc.Natl.Acad.Sci.USA* **83**(1986), 9373-9377.

36. D.T.Gillespie. *J.Comp.Phys.* **22**(1976), 403-434.

37. M.Nowak, P.Schuster. *J.Theor.Biol.*, in press.

38. J.B.Kruskal. *Proc.Amer.Math.Soc.* **7**(1956), 48-50.

Explicit Observers

O.E. Rössler

Institute for Physical and Theoretical Chemistry, University of Tübingen,
D-7400 Tübingen, Fed. Rep. of Germany

Abstract. A new approach to the brain and the world is presented. A chaotic Hamiltonian universe is set up, in 1 D, such that an explicit internal observer – an excitable system – becomes amenable to complete understanding. The Gibbs symmetry and the Wigner symmetry, when taken into explicit regard, imply that to this observer, his world appears quite different from what one would expect at first sight – such as when one is doing a molecular dynamics simulation of the same system, for example. Specifically, both stochastic mechanics and the quantum nonlocality turn out to be formal implications of the present "deterministic local hidden variables" approach to quantum mechanics – despite the fact that it never was an approach to quantum mechanics in the first place. Bell's well-known impossibility theorem is circumvented because all quantum effects arising are nonexistent objectively. They are valid only within the "interface" that develops internally between the observer and his world. For the first time, the Kantian notion that the world is objectively different from the way we perceive it can be demonstrated – not for our own world, but for a lower-level model world as it appears to an artificial observer living inside.

1. Introduction

Artificial life is an approach to evolution and the brain [24,26]. As such it is explicit in the sense that the structures in question can be implemented in principle in the laboratory and/or the computer. However, the pertinent equations are in general dissipative. This means that for virtually every such system, an even "more explicit" description is possible in principle in which the irreversible equations are replaced by an underlying set of reversible ones.

An example is Fredkin's discovery that any nontrivial cellular automaton (CA) can be embedded in a larger reversible one [10]. Only the latter would be "maximally explicit" or fully explicit. Similarly, the ordinary or partial differential eqations (ODEs, PDEs) describing an evolutionary soup [23,24] or system of well-stirred [25] or non-well-stirred neurons (cf. the simulation of an excitable PDE with nontrivial behaviour in [35]) are necessarily dissipative. This time, an underlying, maximally explicit description contains not just twice as many variables as in the CA case [10], but 10^3 to 10^{23} times as many. A fully explicit description would be a molecular dynamics (MD) simulation [1] in which the "macroscopic" behaviour is the outgrowth of a reversible microdynamics of vastly more Hamiltonian variables.

The same macro/micro distinction also applies to virtually every computer built so far. The discrete, finite-state description is nothing but a shorthand that captures certain invariant features of an underlying continuous system of ODE's that is "decomposable" in the mathematical sense [27]. This latter system in turn belongs to the class of dissipative ODE's or PDE's already dicussed [25]. Thus, every computer-implemented observer in our own world will once again require an underlying Hamiltonian description on the microlevel.

Ultimately, of course, any such description would, in our own world, have to be replaced by a description in terms of the Schrödinger equation, as a first step on the way towards a final quantum field-theoretic description hopefully available in the future.

The advantage of the artificial life approach, however, is precisely that nothing forces us to choose an explicit universe that is "completely realistic" at the same time. Thus, artificial observers that arise in a reversible cellular automaton are of the same standing, in principle, as artificial observers arising in a molecular dynamics simulation, and as will be artificial observers generated in a "quantum molecular dynamics" simulation of the future.

In the following, the second category among these (the currently accessible M.D. case) will be considered. All results found will nonetheless apply equally to a lattice automaton analogue (first category).

2. Explicit Dissipative Structures

The observers to be simulated belong to the class of dissipative structures - or, synonomously, far-from-equilibrium structures - which figure so prominently in the work of Prigogine [20]. A decade ago, the prevalent attitude among chemists was that going down to this "next lower" level of description would almost never be necessary when dealing with a chemical system like the Zhabotinsky reaction, for example. Certainly, you would have to make sure that your rate equations are open in the sense of containing pool substances that either are kept constant (so that the system develops an attractor) or are initially far enough away from their final equilibrium values that a nontrivial transient dynamics can develop - as is the case when the Belousov-Zhabotinsky reactants have been freshly poured into a beaker (or when a portable radio has been equipped with a new battery). Even the principle of detailed balance [41] can be incorporated into the macroscopic rate equations, directly.

Of course, it would be "desirable" to include fluctuation terms in the deterministic rate equations under certain conditions - as in the presence of a bifurcation or phase transition - in order to capture the details of the amplified fluctuations that then occur. "In general", however, knowledge that the need for a second-order approximation may arise would suffice.

Until very recently, there was perhaps indeed no need to bother. Only physico-chemical conscientiousness - and some technical applications - seemed to justifiy research programs that from the beginning included the next-lower phenomenological level [20]. From the point of view of efficacy - when the finding of new phenomena like chaotic behaviour on the macro-level was at stake, for example [28] -, it would even be a mistake to keep the next-lower level of description in mind continually. The hierarchy of sciences - quantum field theory at the bottom, sociology at the top - fortunately works most of the time. It would be a major catastrophe if it turned out that to understand the brain, for example, you have to use the quantum level of description.

The chaos paradigm changed this attitude. Suddenly there was a dynamical principle which (a) was very powerful in terms of the predictions it made (which included the presence of an irreducible element of unpredictability [39]) and (b) was independent of the level within which it was applied. Populations of butterflies that each employ a chaos-generator in their individual flight maneuvers could easily behave chaotically, for example. Of course, the mere presence of analogies between levels does not yet mean that the levels themselves have been punctured. Still, chaos theory for the first time offered a unified approach to chance events on all levels. Might, for example, the irreducible stochastic behaviour found on one of those levels - the quantum level - really prove resistant to the new paradigm?

Looked at this way, the Boltzmann-Prigogine research program of going directly to the "correct" level of description when phenomena like entropy increase and evolution are at stake, suddenly made eminent sense. After all, no one had yet fully understood how reversible trajectories generate an irreversible macrodynamics, for example. Chaos theory, applied to systems with a few degrees of freedom, might be able to reveal the ultimate mechanism [30].

While the reversibility paradox thus appeared in a new light, should one really believe that this paradox might have anything to do with quantum mechanics, or with the way an observer perceives his world? Moreover, would this latter possibility not invite back the catastrophe of level mixing?

The answer is yes. Nevertheless, chaos demands that one try. All that is required is to have another look, after Boltzmann and Gibbs, at what "really" happens on the microlevel. The difference would be that while Boltzmann and Gibbs had thought they were doing physics, this would be nothing but chaos theory. Whether or not the lowest level of nature can be expected to comply is of secondary relevance, as mentioned.

At this point, everything becomes very simple. There is this molecular-dynamics simulation paradigm [1,4] which, it turns out, is pure chaos from the beginning (as Maxwell had known). It is a universe to explore.

3. Molecular Dynamics-Born Observers

Dissipative structures have been simulated before from first reversible principles, in an MD (molecular dymanics) approach. The example of bistability (and by implication autocatalysis) was studied in this way [13]. Nevertheless no analogue to the Zhabotinsky reaction (with its autonomous oscillatory and/or chaotic behaviour [14]) appears ever to have been simulated explicitly.

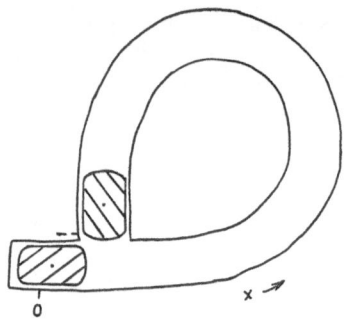

0   x →

Fig. 1:   A chaotic 1-D Hamiltonian system.

The reason for this is simple. The number of particles needed to get a working simulation (in which the autonomous macroscopic behaviour is not obliterated by noise), in 3 or 2 space dimensions, is prohibitively high even for present-day computer power.

Therefore a return to the chaos-theoretic approach is called for. Indeed it turns out that no more than a single dimension is needed. For an example, see at Fig. 1.

Here we have two loaf-shaped particles living in a frictionless 1–D tube. The corresponding Hamiltonian is

$$H = \sum_{i=1}^{n} \frac{p_i^2}{2m} + \varepsilon \sum_{i=1}^{n} \frac{1}{x_i} + \varepsilon \overset{n,n-1}{\underset{i,j=1,2}{P}} \frac{1}{1 - f(x_i) - x_j} \tag{1}$$

Here P is the pair permutation operator, the masses are equal (m=1), the particle number is two (n=2), and f is the blunted-tent-in-the-prairie function, for example:

$$f = \sqrt{(x-0.05)^2 + 5 * 10^{-6}} + \sqrt{(x-0.15)^2 + 5 * 10^{-6}}$$
$$- 2 \cdot \sqrt{(x-0.1)^2 + 10^{-4}} \quad ,$$

see references [30,31,33,36].

The system of Figure 1, described by (1) with n=2, is clearly chaotic. This becomes obvious from Figure 2 which shows the position space (also called configuration space) for this system as ε approaches zero, cf. [30]. One sees that the system is equivalent to a 2-D chaotic Sinai billiard system [40]. Specifically, neighbouring trajectories - impinging in parallel on the two curved protrusions in Fig. 1 - will diverge and will keep doing so exponentially [40]. Therefore there is a positive Lyapunov characteristic exponent - implying chaos - present in this 1-D Hamiltonian system.

Moreover, it is easy to generalize the system to arbitrarily high particle numbers and dimensions without it losing its mathematically transparent character. With three particles, for example, we obtain Fig. 3.

There are now 3 different projections looking like Figure 2. The corresponding Hamiltonian is again (1), but with n = 3. This "building-

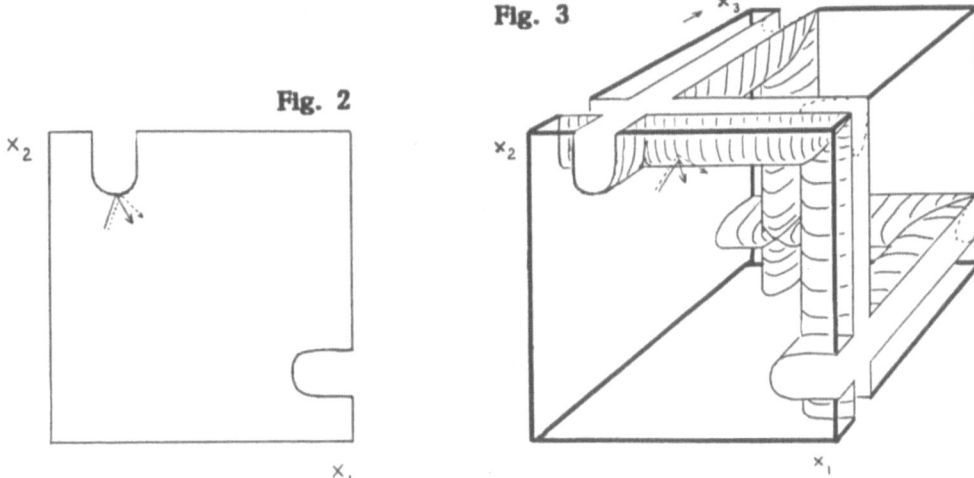

Fig. 2: Configuration space for the system of Figure 1, that is, (1) with n=2 and ε small. Two neighbouring trajectories are shown. See text.

Fig. 3: Configuration space for the system of Fig. 1 and (1) with n = 3. Compare Figure 2.

block principle" continues. For n particles we have n-1 equal positive Lyapunov characteristic exponents ("maximal chaos") [31]. No 3-D gas can be more chaotic [32].

In (1) it is assumed that the interactions are confined to "vertical vs. horizontal" collisions (see Figure 1). This can be guaranteed mechanically by assuming that each "bullet" runs in its own 1-D track, but with this track twisted 90 degrees on the way up from the horizontal to the vertical orientation. On the other hand, no such mechanical "excuse" is really needed in order to render a Hamiltonian function like that of (1) legitimate mathematically.

Now that we have n particles, we can add "colours" to them as well as colour-changing reactions [31]. In this way, a 1-D MD system is obtained in which, with fairly low particle numbers, dissipative structures of the limit-cycle type can be implemented both in the far-from-equilibrium variety and in the "open" (infinitely far from equilibrium) variety [31].

Using the macroscopic rate equations of [31], it is easy to change one or two of the parameters so as to turn the relaxation-type limit cycle indicated in [31] into an "excitable system" (like Winfree's Z reagent [45]). The resulting excitable system will then be an explicit observer.

Of course, it will still be a very crude analogy with a full-fledged "explicit brain". To obtain the latter, not 1 or 10 or 100 such systems would have to be combined (as presently feasible), but 10^{11} or so. Nevertheless, the above excitable system would be a first representative of the whole class.

4. An Explicit Measurement Situation

The excitable system (explicit observer) can be combined with a second explicit dissipative structure which acts as an "amplifier". (Think of a power-brake-like contraption.) The amplifier in turn may be equipped with a light front-end particle that is part of a cooled front-end portion ("pre-amplifier"). Intermittently, a third explicit dissipative structure could be brought into play to do the cooling. The resulting measuring device would than translate the fine motions of an "object particle" into stronger, more or less proportional but slightly delayed motions of its large-mass output particle. The latter would directly impinge on the observer as a mechanical "pointer" [32]. In this way an explicit measurement situation would be obtained.

When simulating this whole system in a computer (which has yet to be done), nothing surprising will happen at first sight. When the pointer pushes strongly enough, macroscopically speaking, the observer will be excited; otherwise it will not. That is all.

The properties just described are exactly those that one would naively expect from any classical microscopic description of a macroscopic observer who uses a macroscopic measuring device to observe a microscopic particle. Such a classical universe is considered completely useless for understanding anything about our own world, which, as is well known, is nondeterministic on the microlevel. Furthermore, even if there were something to be learned about the quantum indeterminacy, there would still be the nonlocal quantum correlations left which as is well known, cannot be reproduced in any standard deterministic classical model according to Bell's theorem [2].

Unexpectedly, while all of this is true, it is also subtly wrong. To be more precise, it is correct on the "exophysical level" (considered so far) but false on the "endophysical" one. Let us now turn to the latter.

5. Endophysics of the Explicit Observer

The fundamental distinction between exo- and endophysics (not to be confused with end-o'-physics) is fairly new, see [32] for a review. The difference is analogous to the distinction introduced by Gödel [12] between generally valid theorems on the one hand and theorems accessible from the inside of the axiomatic system in question on the other. While Gödel's example indeed forms a special case, the question focused on in physics is more restricted: How does the explicit mathematical universe in question appear when it is viewed from the inside? More specifically, there are now two questions: (a) Are there external features left over that are not accessible from the inside? (b) Are there internally valid features that nevertheless are absent when the same system is viewed from the outside?

The latter question was first posed by Maxwell, [17] who showed that to an outside operator (he said "being" or "demon" for lack of a better term), the second law will not exist. Later, Finkelstein [8] in a discrete context questioned the external validity of both relativity and quantum mechanics (see also [9]). The latter of these hypotheses was also introduced in [29] in a continuous context [32].

How does the present explicit world appear to the internal observer? To see this, it is necessary to give the dynamics already considered a "second look". The first to do so, in the context of classical gas at equilibrium, was Gibbs [11]. Unexpectedly, there is a symmetry principle at work. The latter drastically changes the behaviour even after one had already thought there was nothing left to be learned about the system. Gibbs' symmetry is implicit in (1) above. The different particles were assumed to be equal in all physical and mathematical properties (except for position and momentum) as long as they belong to the same class including colour.

Gibbs showed that under this condition, phase space volume has to be divided by N! in the simplest case (when there are N equal particles of the same class present) and by $N_1! \cdot N_2! \cdot ...$ in general [11].

The result was later shown to be mathematically sound when Weyl [42] rediscovered it independently. Weyl's idea was recently worked out [34]. Qualitatively, it all goes back to Leibniz [15], see [42,34].

The essence of the principle is as follows. Particle indistinguishability (in the sense of their being mathematically equal like solitons [46]) implies trajectorial multi-uniqueness. This is a new mathematical feature of certain systems of ordinary differential equations including most Hamiltonian systems. The new axiom has new mathematical implications [32,34].

In short, the system in question becomes confined to a subcell of its phase space - one of N! many identical copies that together make up the full phase space. This group-theoretic reduction [32] is merely a matter of descriptive economy as long as it is the whole universe (observer + measuring device + object) that is considered. However, when for some reason one is interested in the relationship between a subsystem and the rest of the universe, new features suddenly arise. For example, the phase-space volume of a classical gas still goes down by a factor of N!, even if the whole universe into which it is embedded is not considered simultaneously. This finding has measurable consequences for the equilibrium entropy of the gas (which is proportional to the logarithm of phase-space volume [11]. These consequences are obeyed by nature [37].

Similar findings hold for the subsystem of interest in the present case, the explicit observer. This is perhaps best seen with the aid of an example. Assume that the observer consists of but two identical particles (both red), with a third particle (blue) acting as the "rest of the world". Then the system is exophysically described by (1) with n = 3, with Figure 3 describing its position space. Trajectorial bi-uniqueness then has the consequence that phase-space volume becomes halved (Figure 4). At the same time a new, "endophysically correct" Hamiltonian replaces (1). It reads

$$H^* = H + \frac{\varepsilon'}{x_2 - x_1} , \qquad (2)$$

where H is taken from (1) and $\varepsilon' \to 0$ [32].

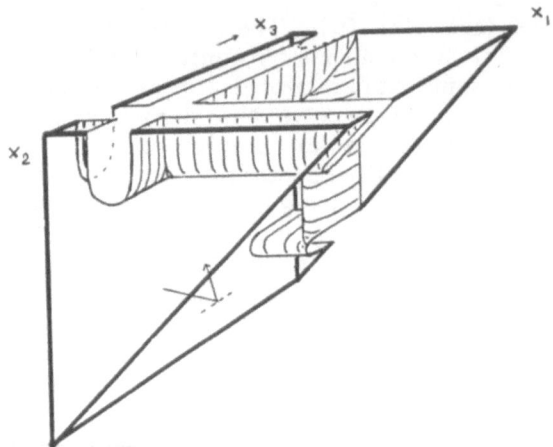

Fig. 4: Halved (reduced) configuration space for the system of Fig. 3, valid in the presence of an exchange symmetry between two of the three particles.

The blue particle no longer interacts with equal probability with both red particles - as it would in the absence of the symmetry - , but only with the first one. Note that according to (2), the first red particle is now confined to the lower values of x while the second preferentially occupies the higher values. This is because the two particles cannot pass by each other any more, endophysically, since they necessarily exchange their identities whenever they pass by each other [32,34].

This can also be seen geometrically. In Figure 4, only one vertical bar (in the back of the cell) remains. It corresponds to the interaction between x_3 and x_1 (see Figure 3). The symmetric interaction between x_1 and x_3 (see the other vertical bar in the foreground on the right in figure 3) now belongs to the other cell.

In general, the new Hamiltonian contains N-1 terms of the type present in (2), namely,

$$H^* = H + \varepsilon' \left(\frac{1}{x_N - x_{N-1}} + \frac{1}{x_{N-1} - x_{N-2}} + \dots + \frac{1}{x_2 - x_1} \right) \qquad (3)$$

if there are N equal particles present [31].

What does this volume-reduction mean for the observer?

6. Implications

It could be for example that the new dynamics is, while quantitatively different, nevertheless qualitatively still the same as the old dynamics. Even this, however, is not the case. While the old dynamics was maximally chaotic, the new dynamics is essentially nonchaotic, namely, quasiperiodic [31,33]. The reason lies in the fact that the equal particles can no longer pass by each other. Most "inner" paticles in the ring of Figure

I therefore no longer reach the point where the action (chaos generation) takes place.

The new Hamiltonian (3) clearly generates different trajectories to those of the original one (1). Only locally is there still coincidence with one of the original (multi-unique) trajectories [32]. A similar result still holds true in 2- and 3-D Hamiltonians ([32] and work in preparation).

Endowed with this main result, we can now ask for its implications. Twelve were recently described [32]. The main implication was that any directly coupled external object will be subject to an effective diffusion coefficient, of magnitude

$$D = \frac{\varepsilon \, \tau}{2 \, M} \quad . \tag{4}$$

Here M is the object's mass, τ is the mean cell passage time inside the observer, and $\varepsilon = 1/(2kT)$ is the mean energy per particle of the observer. A directly coupled external object therefore performs a Brownian motion determined by D relative to the observer. This is not too surprising since apart from the thermal noise, ε , only τ enters, a characteristic period arising within the observer. The noise energy, ε , is effectively reapplied every unit time interval, τ , to the external object. This leads directly to (4) [32].

Unexpectedly, the same result, (4), applies, to "indirectly coupled" objects as well. Indirectly coupled objects are, for example, objects, that are observed by the observer as the latter makes use of a measuring apparatus [32].

This latter result, if correct, is surprising. The reason is that (4) with $h/2\pi$ standing for $\varepsilon\tau$, is well known in quantum mechanics, where it is called Nelson's postulate. Nelson [18] postulated this equation as an axiom and showed that this single assumption suffices to obtain the Schrödinger equation and most of quantum mechanics [38]. Thus, Nelson's stochastic mechanics (and therefore quantum mechanics) would be an implication of the present classical universe.

The question of whether (4), with $(h/2\pi)_{obs} \equiv \varepsilon\tau$ replacing $h/2\pi$, is indeed an endophysical implication of (1) is, therefore, the main question (no. 1) to be considered in the following. A second question immediately follows. (2) If stochastic mechanics follows from first principles, how about the quantum nonlocality? (Nelson [19] so far only showed the compatibility-in-principle between his theory and nonlocality.) Thereafter would come the following two questions as a coda. (3) How can we reconcile the facts that the large-mass pointer suffers no uncertainty, while the low-mass object whose motion it proportionally represents does? And (4) How does the fact that $(h/2\pi)_{obs}$ is not an objective constant but a constant only for the observer, make itself felt to the observer?

7. Main New Principle

The major result underlying the above approach was "trajectorial N!-uniqueness" in the presence of particle indistinguishability. This result, it turns out, is closely related to an even more general principle: "trajectorial bi-uniqueness" in <u>all</u> Hamiltonian systems of the Newtonian type. This latter result, to be demonstrated now, will then provide answers to the four questions posed above.

Recall the role of the "quasiperiod" τ. The observer is effectively subjected to an oscillation of that period [32]. But why should that fact greatly affect his observations? The reason was that every typical motion inside the observer, and every external motion, reverse their mutual orientations every unit time interval τ. Therefore, so the reasoning went [32], the relative relation is the same as if the observer did not change his orientation in time every unit time interval τ, but rather the external particle (and the whole measuring chain) did.

This reasoning appears flawless. If time does not flow (as it is the case in science, see [5]), the two situations must be equivalent. On the other hand, the question of time reversal invariance, which is thereby elevated to unexpected prominence, certainly deserves to be re-addressed in its own right before the final verdict.

The new group-theoretic instrument (analysis of multi-unique trajectories [32]) provides a solution to this new question as well. In fact, the main point was already recognized by Wigner [44] in the context of quantum mechanics. Only later did the evidence accumulate that an analogous treatment can be applied to classical systems [16,16a].

Let us adapt the results to the present context. Hamiltonian systems in general possess two equivalent trajectories instead of a single one (a "black" and a "white" one for short). The <u>black</u> trajectory is the one usually considered. Specifically, given a Hamiltonian H quadratic in p, Hamilton's recipe yields,

$$dx/dt = p \quad (= dH/dp) ,$$
$$dp/dt = f \quad (= - dH/dx) ,$$

(5)

where f is the Newtonian force. The initial conditions are $x_0 = x_0$ (the given initial position) and $p_0 = mv_0$ where v_0 is the initial rate of change of x ($v_0 = dx_0/dt$).

However, there is a second, equivalent trajectory implied by H. This <u>white</u> trajectory is defined by

$$dx*/dt = -p* \quad (= - dH/dp*) ,$$
$$dp*/dt = - f \quad (= + dH/dx*) ,$$

(6)

with initial conditions $x*_0 = x_0$ and $p*_0 = - p_0$. The bijective transformation rule leading from (5) to (6) is $x* = x$ and $p* = - p$.

Equation (6) appears to be equivalent to (5) under time reversal. Inverting time ($t* = - t$) in (6), one obtains

$$dx*/dt = p* \quad (= dH/dp*) ,$$
$$dp*/dt = f \quad (= - dH/dx*) ,$$

(6')

with $x*_0 = x_0$ and $p*_0 = - p_0$ as before (since only the direction of flow, and not the initial conditions, are changed under mathematical time reversal, $t* = - t$). As expected, (6') is identical to (5) if both x and x* and p and p* are identified. However, there remains an unexpected difference: The initial conditions are different. They were x_0 and p_0 in (5) and are x_0 and $-p_0$ in (6'). The former pair determines the "black" trajectory in both directions of time, the latter pair determines the "white" trajectory in both directions of time. Thus, there indeed exist two trajectories in general.

Wigner [43] said it all in one sentence. "Physical time reversal is motion reversal"; that is, both the direction of the time and the signs of all velocities have to be changed simultaneously. This sentence means that Hamiltonian systems in general possess two trajectories in their phase spaces. They are mirror images to one another, with the one obtained from the other through reflection at the point $p_i = 0$ (all momenta equal to zero). This much is well known in principle [16a].

What is new is that this is nothing but another instance of trajectorial bi-uniqueness. Accordingly, the exophysically correct phase space collapses down to half its former volume once more: In the "reduced" phase space, trajectorial uniqueness is re-established [32]. The new unique trajectory differs from each of its predecessors in that it consists of alternating black and white pieces connected by zero-duration "jump segments" [32].

This however, is exactly the needed piece of information that had been missing. It had been used implicitly, but not explicitly, in the derivation of (4). The "black" and "white" time slices that imply (4) [32] indeed exist. This completes the answer to question (1) in sect. 6.

Let us now turn to question (2).

8. Nonlocality

It goes without saying (but is nevertheless important to realize) that the Brownian motion described by (4) is not something that exists objectively (exophysically) in the world assumed. Rather, it exists by definition only for the observer (endophysically). Nevertheless it represents inescapable reality to the observer. It is a property of the "interface" between the observer and his world; in this sense it is universal.

Now, the very non-objective validity of $(h/2\pi)_{obs}$ is the reason why its implications are not covered by Bell's [2] theorem which states that deterministic hidden variables of the classical local type can never reproduce the quantum nonlocality. Thus, if the present theory turned out to imply quantum nonlocality, this "endo" counter example would not violate the original ("exo") theory.

To see whether this is indeed the case, consider two exophysically correlated objects (of zero combined angular momenta, for example) existing in the model world. Since the observer generates all perturbations through his own passing through black and white time slices in a particular fashion, it follows that both objects will be perturbed identically. Therefore a subsequent deviating measurement, made on one of the two particles, will show the same outcome as if the first measurement had been performed on this same particle even if it were not. Hence the unilaterally valid $(\cos^2)$ law [38] applies also in the bilateral case. From this fact follows a violation of Bell's inequality [2].

Thus, as well as stochastic mechanics, the quantum nonlocality is also an implication of the present deterministic local hidden variables universe.

9. Discussion

A microscopic view of a conceptually transparent universe has been presented. This universe was chaotic and classical. The implications, nevertheless, were nonchaotic and nonclassical. Moreover, for the possibility arose the first time that microscopic features may percolate all the way up to affect the macroscopic appearance of the world. Physics becomes dependent on brain theory.

This is surprising enough to warrant a critical second look. What is the weakest point in the proof? Probably, it lies not so much with what has been shown above as with what has not yet been shown; namely, how can a non-perturbed massive pointer mediate perturbed behaviour of a light object particle when the two are actually strongly correlated? This question, (3) above, is much harder to answer than were its two antecedents (stochasticity and nonlocality) considered in the preceding sections.

There is hope that the question can eventually be answered using the "x-t-plot" technique (all particle position vs. time). The latter contains all information and indeed allowed the first spotting of the time sclices [32]. However, the right way of compressing and processing this information so that it makes intuitive sense again remains to be found. Therefore, only a qualitative argument can be offered for the time being.

The above contradiction can be resolved only if the pointer particle is not registered objectively as long as it is part of the measuring chain (which turns black and white in synchrony with the observer [32]). On the other hand, the same particle must appear normal when taken as an object in its own right. Niels Bohr [6] once pondered a related question when claiming that a blind man can only either feel _with_ the cane or feel _the_ cane. Accomplishing both things simultaneously was considered impossible. This, however, is precisely the task required here.

The following possibility comes to mind. The pointer must effectively "freeze out" in order to represent the presently valid perturbed behaviour of the object. At the next moment, with a different sequence of black and white time slices in charge, the object will have to appear perturbed differently. The pointer will in that case have to appear frozen out differently; and so on at every moment. In other words, the whole macroscopic world (including the pointer and its past) must appear consistent in its supporting a certain perturbed state of the object at every moment, but it must do so differently at every moment. This implies that the change involved is not accessible to the observer. Thus, a "masking principle", shielding the observer from inconsistency-generating information, is invoked.

134

This "desperate" proposal may or may not be correct (verifiable). Unexpectedly, it at the same time represents a solution to the measurement problem of quantum mechanics. It is closely related to two recent interpretations put forward by Bell [3] and Deutsch [5a] (cf. also [21]) within the context of Everett's [7] so called many-worlds interpretation of quantum mechanics. At every moment, a different Everett universe is in charge without this fact being accessible to the observer [3]. "Nowness" becomes codimension-two (there is a second free parameter) rather than merely codimension-one (a point in the time axis) as one would usually think [7, 21, 5a]. Thus Everett's version of quantum mechanics is effectively implied by the above proposal. Simultaneously, the remaining problem (4) is also automatically taken care of by the masking principle.

Thus, the three major riddles of quantum mechanics (stochasticity; nonlocality; state reduction) all re-surface in the present artificial universe, either as an implication (the first two) or as a proposed implication (the third). A new fourth problem also arises but turns out to be a special case of the third. All of these implications appear to be typical, not just for the present reversible universe, but for "most" revesible universes.

There may be further major implications of reversible universes. Let us, however, turn to a more general question. Suppose an artificial world like the one described above were actually built some day. Suppose further that developments in computer technology make it possible to implement many excitable systems (formal neurons) simultaneously so that an arbitrarily complicated reversible artificial brain (or several) can be built. Will the inhabitants be able to find out? A similar question was first posed in a dissipative context by Putman [22]. He convincingly argued that the answer must be negative. In the present reversible context, a positive answer appears to be possible.

Let me finish by describing a "meta-experiment" which the inhabitants would be able to perform after having embarked on an endophysical program of their own. The experiment is called the "fever test".

The observer first calculates his own $(h/2\pi)_{obs}$ on the basis of available information (mass density of the smallest particles that make him up; body temperature). Then he compares this value with $(h/2\pi)_{univ}$ from the books. If the two coincide to one digit, he tries to improve on the result by measuring and calculating more accurately until the coincidence becomes very good. (This would obviously be possible in the model world). Now begins the second, decisive phase. He takes a fever pill and re-does the calculation with the new data collected while he has a fever. He predictably finds a deviation in the fourth digit between his own $(h/2\pi)_{obs}$ and that of his friends who do not have a fever. Then he again compares with the books. The experiment is rated a success if one of the calculated numbers consistently turns out to be privileged - his own.

In the model world, this activity would actually be complicated by a "sociological" problem. While the experimenter could easily perform the experiment as described, he would have great difficulty convincing others of its seriosity. Even the first step - the single-digit coincidence assumed for starters - would be considered an unpublishable observation by his colleagues. In their nonrelativistic world, there would never have been any reason (like superluminal [2] nonlocality) to question the quantum laws' objective reality. The endophysical insight that interfaces are also something objective would come less easy. Moreover, our own world differs from theirs in an additional respect. So many important forces (radiation, gravity, nuclear) were omitted in our setting up the artificial world that the latter does not qualify as a model. This allows the exotic fever test appear acceptable to us.

To conclude, reversible worlds are different. Owing to the Wigner symmetry, external causality vacillates for every subsystem. The implications regarding observation are bound to be far-reaching. So far, only some fairly tame effects resembling quantum mechanics could be pinpointed. Nevertheless Parmenides' and Kant's assertion that reality is different from the way it appears can already be confirmed – in a toy setting. The science of AL (artificial life) can do more than open up new worlds for us. It may re-define our own.

10. Acknowledgement

Invited paper, presented under the title "Explicit Observers in an Invertible Artificial World", at the "Artificial Life Workshop", Set. 21-25, 1987, Center for Non-linear Studies, Los Alamos, Sept. 21, 1987.
I thank Chris Langton and Gottfried Mayer-Kress for stimulation. Conversations with Norman Packard, Doyne Farmer, Jim Crutchfield, Jeffrey Tennyson and Dietrich Hoffmann were also very helpful.

11. References

1 B.J. Alder, and T.E. Wainwright, "Phase Transitions for a Hard Sphere System", J. Chem. Phys. **27** , 1208 (1957).

2 J.S. Bell, "On the Einstein-Podolsky-Rosen Paradox", Physics **1,** 195 (1964).

3 J.S. Bell, "Quantum Mechanics for Cosmologists", Quantum Gravity Vol. **2**, edits. C.J. Isham, R. Penrose, and D.W. Sciama, Oxford Univesity Press, (1981) p. 611

4 B.J. Berne, Statistical Mechanics, Part B: Time-Dependent Processes, Plenum Press, New York (1977).

5 P.C.W. Davies, The Physics of Time Asymmetry, London, Surrey University Press (1974).

5a D. Deutsch, "The Connection between Everett's Interpretation and Experiment", Quantum Concepts in Space and Time, edits. R. Penrose and C.J. Isham, Oxford, Clarendon (1986) p. 215

6 P.A.M. Dirac, "The Versatility of Niels Bohr", Niels Bohr, edit. S. Rozental, Elsevier, Amsterdam (1967), p. 306.

7 H. Everett, "Relative State Formulation of Quantum Mechanics", Rev. Mod. Phys. **29**, 454 (1957).

8 D. Finkelstein, "The Holistic Methods in Quantum Logic". Quantum Theory and the Structures of Time and Space Volume III, edits. L. Castell and C.F. Weizsäcker, Carl Hanser Verl., München (1979) p.37.

9 D. Finkelstein and S.R. Finkelstein, "Computer Interactivity Simulates Quantum Complementarity", Int. J. Theor. Phys. **22**, 753 (1983).

10 E. Fredkin and T. Toffoli, "Conservative Logic", Int. J. Theor. Phys. **21**, 219 (1982).

11 J.W. Gibbs, Elementary Principles in Statistical Mechanics, CT: Yale University Press, New Haven, (1902) ch.15.

12 K. Gödel, On Formally Undecidable Propositions, Basic Books, New York (1962), (originally published in 1931).

13 M. Heinrichs and F.W. Schneider, "Molecular Dynamics Calculations of a Second-Order Kinetic Phase Transition in an Open System (CSTR)" Ber. Bunsenges. Phys. Chem. **87**, 1195 (1983).

14 J.L. Hudson and O.E. Rössler, "Chaos and Complex Oscillations in Stirred Chemical Reactors", Dynamics of Nonlinear Systems, edit. Vladimir Hlavacek, Gordon and Breach, New York (1986) p.193.

15 G.W. Leibniz, The Leibniz Clark Correspondence, edit. H.G. Alexander, Manchester University Press Barnes and Noble, Manchester (1956) pp. 26, 38, and 63.

16 G. Lüders, "On Motion Reversal in Quantized Field Theories" (in German) Z. Pysik **133**, 325 (1952).

16a R.L. Devaney, "Reversible Endomorphisms and Flows", Trans. Amer. Math. Soc., **218**, 89 (1976).

17 J.C. Maxwell, Theory of Heat, Appleton, New York (1872) p. 309.

18 E. Nelson, "Derivation of the Schrödinger Equation from Newtonian Mechanics", Phys. Rev. **150**, 1079 (1966).

19 E. Nelson, "The Locality Problem in Stochastic Mechanics", Ann. N.Y. Acad. Sci. **480**, 533 (1987).

20 G. Nicolis, and I, Prigogine, The Investigation of the Complex, (in German), Piper Verl., München (1987).

21 D. Page, and W.K. Wootters, "Evolution without Evolution - Physics Described by Stationary Variables", Phys. Rev. **27D**, 2885 (1983).

22 H. Putman, Reason, Truth and History, Cambridge University Press, Cambridge, (1981), ch. 1.

23 O.E. Rössler, "A System-Theoretic Model of Biogenesis" (in German) Z, Naturforsch. **26b**, 741 (1971).

24 O.E. Rössler, "Design of Autonomous Chemical Growth under Different Environmental Constraints", Progr. Theor. Biol. **2**, 167 (1972).

25 O.E. Rössler, "A Synthetic Approach to Exotic Kinetics (with Examples)", Lect. Not. Biomath. **4**, 546 (1974).

26 O.E. Rössler, "Adequate Locomotion Strategies for an Abstract Organism in an Abstract Environment: A Relational Approach to Brain Function", Lect. Not. Biomath. **4**, 342 (1974).

27 O.E. Rössler, "Chemical Automata in Homogeneous and Reaction-Diffusion Kinetics", Lect. Not. Biomath. **4**, 399 (1974).

28 O.E. Rössler, "Chaotic Behaviour in Simple Reaction Systems", Z. Naturforsch. **31a**, 259 (1976).

29 O.E. Rössler, "Chaos and Chemistry", Nonlinear Phenomena in Chemical Dynamics, edits. C. Vidal and A. Pacault, Springer Verl. New York, Heidelberg, (1981) p. 79.

30 O.E. Rössler, "Macroscopic Behaviour in a Simple Chaotic Hamiltonian System", Lect. Not. Phys. **179**, 67 (1983).

31 O.E. Rössler, "A Chaotic 1-D Gas: Some Implications", Lect. Not. Phys. **278**, 9 (1987).

32 O.E. Rössler, "Endophysics", Real Brains, Artificial Minds, edits. John l. Casti and Anders Karlqvist, North-Holland Publ., New York, Amsterdam, (1987), p. 25.

33 O.E. Rössler, "Explicit Dissipative Structures", Found. Phys. **17**, 679 (1987).

34 O.E. Rössler, "Symmetry-Induced Disappearance of Reality: The Leibniz Effect", Leonardo **21**,4 (1988), Special Issue on Art and the New Biology: Biological Forms and Patterns, edit. Peter Erdi, December.

35 O.E. Rössler and C. Kahlert, "Winfree Meandering in a 2-Dimensional 2-Variable Excitable Medium", Z. Naturforsch. **34 a**, 565 (1979).

36 O.E. Rössler and M. Hoffmann, "Quasiperiodization in Classical Hyperchaos", J. Comp. Chem. **8**, 510 (1987).

37 O. Sackur, "Applying the Kinetic Theory of Gases to Chemical Problems", (in German), Ann. der Phys. **36**, 958 (1911).

38 W.R. Schneider, "Stochastic Mechanics", Quantum Mechanics Today, Lect. Not. 11th Gwatt Workshop, October 15-17, (1987), edits. D. Baeriswyl, M. Droz, C.P. Enz and A. Malaspinas (B.B.C, CH-5405 Baden), p. 234.

39 R. Shaw, "Strange Attractors, Chaotic Behaviour and Information Flow", Z. Naturforsch. **36 a**, 80 (1981).

40 Ya.G. Sinai, "Dynamical Systems with Elastic Reflections", Russian Math. Surveys **25**, 137 (1970).

41 R. Wegscheider, "On Simultaneous Equilibria and the Relations Between the Thermodynamics and the Reaction Kinetics of Homogeneous Systems" (in German), Z. Phys. Chem. **39**, 257 (1902).

42 H. Weyl, Philosophy of Mathematics and Science, Princeton University Press, Princeton, N.J. (1949), ch. 22 e, append. B.3.

43 E.P. Wigner, "Relativistic Invariance and Quantum Phenomena", Rev. Mod. Phys. **29**, 255 (1957).

44 E.P. Wigner, Group Theory and its Application to the Quantum Mechanics of Atomic Spectra, Academic Press, New York, (1959) ch. 26; (Material originally published in 1932).

45 A.T. Winfree, The Geometry of Biological Time, Springer Verl., New York, Heidelberg, (1980) p. 240.

46 Solitons (particle-like solutions) that arise in 2-D Hamiltonian media indeed come in discrete classes so that the present theory is applicable (Richard Bagley, private communication 1987).

Dynamics of Networks and Pattern Processing

W. Ebeling, I. Schimansky-Geier, and Ch. Zülicke

Humboldt University of Berlin, Department of Physics,
Invalidenstraße 42, DDR-1040 Berlin, German Democratic Republic

1. Introduction

The dynamics of networks is closely connected with the development of the modern science of non-linear irreversible processes [1 - 6].

Let us begin by introducing several concepts taken from the field of statistical physics and thermodynamics. We firstly have to deal with the relation of entropy and information [7, 8].

All processes in networks, and particularly in information processing, involve changes in entropy. The close connection between entropy and information follows from the relation between Shannon entropy and Boltzmann-Gibbs entropy. For the purposes of information theory we describe, in the same manner as Shannon, the state of a system in terms of a set of macroscopic state variables X (macroscopic state space). Assuming the probability distribution $P(X)$ the Shannon entropy of the system is defined by

$$H = - \int dX \, P(X) \, \ln P(X).$$
(1.1)

Here, X is to be understood as a complete set of order parameters describing the macroscopic state relevant for the observation or information transfer process. On the other hand, the Boltzmann-Gibbs entropy refers to a special state space, the phase space of the microscopic coordinates and momenta (in the classical theory) or the Hilbert space of the quantum states (in the quantum theory). Let us for simplicity consider only the classical case of an N particle system with the phase points

$$x = (q_1 \ldots q_N \, p_1 \ldots p_N)$$

and the normalized probability distribution

$$P(x) = P(q_1 \ldots q_N \, p_1 \ldots p_N).$$

The Boltzmann-Gibbs entropy is then given by the well-known formula

$$S = - k_B \int dx \, P(x) \, \ln P(x)$$
(1.2)

where k_B is the Boltzmann constant.

The two expressions are evidently identical (up to the constant k_B) in the event that the macroscopic and the microscopic descriptions coincide. Further we have the inequality

$$H \leq k_B^{-1} S .$$ (1.3)

This is based on the fact that the macroscopic description is always a reduced one. The finest possible description of the state of the system is the microscopic one.

Information exchange between two systems is always connected with changes in the Shannon entropy which are accompanied by corresponding changes in the Boltzmann-Gibbs entropy. Due to the basic inequality given above we find a corresponding inequality for the changes

$$\Delta H \leq k_B^{-1} \Delta S$$ (1.4)

We shall now give an interpretation of this important relation. Information transfer always involves entropy transfer. However, the opposite is not true. Information transfer appears to be a special form of entropy transfer. There are other forms of entropy transfer, such as heat conduction, which have nothing to do with information transfer. In our view, inside a body entropy has no form, but is just plain entropy. However, passing from one body to another it assumes a form and one of the possible forms is Shannon entropy, i.e. entropy possessing information character.

Due to the inequalities discussed above, information transfer is bound to the second law of thermodynamics. This law states that entropy can be produced but never destroyed. As formulated by Prigogine, this reads

$$dS = d_i S + d_e S; \qquad d_i S \geq 0 .$$ (1.5)

Here, $d_e S$ denotes the exchange part of the entropy change and $d_i S$ the production part, which is always non-negative according to the second law. In equilibrium systems entropy is maximal i.e. uncertainy is maximal and this excludes information transfer. Decrease of uncertainty is necessarily connected with deviations from equilibrium. Information processing requires non-equilibrium conditions, and in general this is possible only under the condition of a permanent export of entropy to the surroundings. In this way information processing appears as a special form of self-organization of a system. Information processing is possible only in systems which are pumped with high-valued energy. This may be electrical energy in the case of present-day computers, or food in the case of animals.

Let us summarize again: information processing is a special form of self-organization. It requires non-equilibrium conditions and pumping of the system with high-valued energy (which is equivalent to the export of entropy). Information processing involves the transfer of a special form of entropy, called information entropy, from one system to another or the change of other forms of entropy into information entropy. The second law refers only to the sum of all entropy forms ,while the first law refers to the sum of all energy forms only. So far the relationship between thermodynamics and information theory is still unclear in spite of several new contributions to the discussion [7-11]. Evidently the thermodynamics of information processing is still in an initial state of development. At the same time all living systems and technical devices constitute systems in which thermodynamics and information processes play the decisive role. This underlines the urgency of the continuation of the discussion on this matter. Our point of view, developed elsewhere in more detail [7], is the following: the second law refers only to the total entropy of a system. In the process of exchange between sub-systems entropy may assume several forms, one of them being information entropy. The forms of entropy can have an information value if they are exchanged between information processing systems. The quantity of information contained in a message is restricted by the inequality (1.4). The quality of the information contained in a message is specific and cannot be measured in terms of physical quantities. On the other hand, some methods developed by the theory of self-organization (synergetics) are useful in determining the "information content" (see Haken in [9] and [10]). Our hypothesis is that there exists an objective quantity which measures the information quality of entropy in the same way as entropy measures the work quality of energy. So far no method has been found with which to specify this quantity in clear terms.

2. Networks with Double Dynamics

Gustav Robert Kirchhoff, who formulated the current and voltage rules for electrical networks in Königsberg and Berlin between 1845 and 1847 and the general dynamics of currents in conductors in 1857 may be considered the father of network theory. Since Kirchhoff's time, linear network theory has been extensively investigated within the framework of electrical engineering and systems theory. The theory of non-linear networks, however, was developed only in recent decades. The increasing interest in such systems is closely linked to the development of new disciplines such as synergetics [1] and the theory of self-organization [2,3]. There exist many important applications of non-linear networks e.g. to problems of biogenesis [4] and ecology [5,6]. Only recently has it become clear that non-linear networks are also capable of information processing [7-16].

The central point of this paper is that, with regard to application, most interest should be devoted to excitable media [17], and, further, that the basic theoretical concept should be double dynamics [18]. On the other hand, all biological information processing systems are based on the functioning of neural networks which may be thought of as highly non-linear networks. The highest product of natural evolution is doubtlessly the brain of mammals, and the human brain in particular. Information processing by brains is based on the self-organization of billions of neurons which form a dynamic network of very high connectivity and characterized by strong non-linearity.

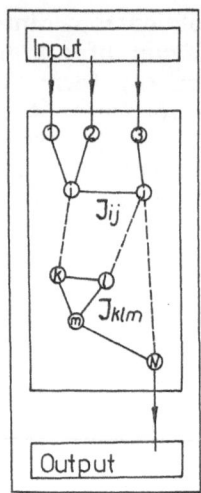

Fig. 1: Graph of the dynamic network of a system.

Let us now introduce various mathematical terms with which to describe networks using graph-theoretical methods [19,20]. In the following we define dynamic networks as a system of numbered nodes i = 1, 2, 3, ..., N connected by binary edges (i,j), ternary edges (i,j,k) etc. The nodes of the edges are associated with, in general, time-dependent vectors $x(t)$ and matrices $J(t)$ respectively, which are real or integer numbers (Fig. 1).

$$x(t) = (\ x_i(t)) \tag{2.1}$$

$$J(t) = (\ J_{ij}(t),\ J_{klm}(t),\ ...\ )\ .$$

The in general non-linear dynamics are determined by a set of parameters, a subset of which is time-dependent, and the remainder fixed, $p(t) = (\ w(t),\ v)$. The dynamics may be introduced e.g. by the map

$$x(t + \Delta t) = T\ (\ x(t),\ w(t),\ v,\ \Delta t) \tag{2.2}$$

$$w(t + \Delta t) = \varepsilon\ U(\ x(t),\ w(t),\ v,\ \Delta t) \tag{2.3}$$

where v are fixed parameters and Δt is infinitesimal or finite corresponding to continuous or discrete mappings. The first part of the dynamics which refers to the state space (2.2) is in general much faster than the second part which refers to the parameters and especially to the couplings (2.3). We may consider ε as a small parameter. This is the concept of double dynamics [18].

The enormous flexibility and adaptibility of dynamic networks of these types is responsible for the vital role they play in evolution. It is just this property which is also responsible for the ability of networks to process information [13-16]. Several information processing functions can now be managed by adaptive networks. One such prototype is the so-called learning machine [21]. Such learning machines have been success-fully applied to problems such as encoding-decoding, or the identification of a word in a given sequence [22], as well as many other recognition problems [23].

Learning machines are characterized by some general principles. Let us briefly discuss them within the framework of a network of non-linear elements.

1) Three functional levels of the network are distinguished: Input $(x^{(i)})$, hidden $(x^{(h)})$ and output $(x^{(o)})$ elements. The aim of the learning ma-chine is to realize a set of certain input-output relations whereby the input values can be given with a certain degree of error.

$$x^{(i)} + \delta x \;\; \rightarrow \;\; x^{(o)} \; . \tag{2.4}$$

During this process the hidden elements play the role of a black box.

2) During the evolution on a quick time scale due to (2.2), dissipative dynamic effects play the essential role: they compress the information in the given input-values into an attractor of the whole network. This at-tractor has a finite region of attraction to allow some perturbations of the input values around a mean value. Thus, several slightly perturbated initial data realize the same output.

3) To realize different input-output relations the system as the who-le has to possess a large number of different attractors.

$$x_1^{(i)} + \delta x \;\; \rightarrow \;\; x_1^{(o)}$$

$$\tag{2.5}$$

$$...$$

$$x_I^{(i)} + \delta x \;\; \rightarrow \;\; x_0^{(o)} \; .$$

For sufficiently separated input data different output values have to occur.

4) For a special set of input-output relations, the network has to be adapted by changing some of the parameters $w \in p$ on a slow time scale due to (2.3) with $\varepsilon \ll 1$. In this approach of double dynamics [18], the attractor landscape is changed in such a way that the quick dynamics T realizes the desired input-output relations. This adaptation can be achieved by optimizing a function G describing the quality of the learning machines, i.e. the accuracy of the realization of the desired input-output relations.

The maps (2.2) and (2.3) may be deterministic or stochastic. An example of a stochastic map with continuous state space is [24]

$$d x (t)/dt = f(x,w,v) + \xi(t) \tag{2.6}$$

with $\xi(t)$ being a stochastic Langevin term. Another important case corresponds to an occupation number space i.e. $x(t)$ are non-negative integers. Then the dynamics is described by master equations [25,26] for the probability distribution $P(x, t)$.

An example for adaptation dynamics due to (2.3) is the gradient system

$$dw(t)/dt = - c \, grad_w \, V(w,x,v) \tag{2.7}$$

optimizing the scalar function V by changing the structure matrix by steepest descent. We speak about learning networks if the system has to react to repeating stimuli (external fields, inputs etc.) and manages somehow to optimize the reaction or output of the system by its dynamics. A useful algorithm for the adaptation was proposed by Ackley, Hinton and Sejnowski [21]. Changing the linking matrix $w = J = (J_{ij})$, they minimize the gain of information G for such a system

$$dw(t)/dt = - \varepsilon \, \partial G/\partial w \; . \tag{2.8}$$

Let us now give some examples of dynamic networks where optimization processes are of great importance.

(i) Mechanical networks [3]

Polypeptides (proteins) and polynucleotides (DNA and RNA) may be considered as mechanical (quantum-mechanical) networks of atoms. The i-th atom is situated at the position q_i and its dynamics are described by Newton's second law

$$m \, d^2q_i/dt^2 + \gamma \, dq_i/dt + \partial U/\partial q_i = \xi_i(t) \tag{2.9}$$

or the overdamped case

$$dq_i/dt = - c \, (\partial U/\partial q_i) + \xi_i(t) \tag{2.10}$$

The potential

$$U(q) = \sum_{ij} \Phi^{(2)}(q_i, q_j) + \sum_{ijk} \Phi^{(3)}(q_i, q_j, q_k) + \ldots \tag{2.11}$$

is minimized by the dynamics (2.9,10). In the course of evolution the macromolecules were also optimized with respect to certain biological functions, e.g. the rate of biochemical reactions.

The dynamics of (2.10) is associated with the probability distribution $P(q, t)$, which obeys the Fokker-Planck equation

$$\partial_t P = \sum_i \partial_i [c \, P \, \partial_i U + D \, \partial_i P] \tag{2.12}$$

where D is the noise intensity of the stochastic forces $\xi_i(t)$ and ∂_i means the partial derivative with respect to q_i. Time-dependent solutions are available from the Schrödinger eigenvalue problem [27]

$$D \sum_i \partial_i^2 \Psi_k(q) + [\varepsilon_k - V(q)] \Psi_k (q) = 0 \qquad (2.13)$$

with

$$V(q) = (\sum_i \partial_i U)^2 / 4D - (\sum_i \partial_i^2 U) / 2 \qquad (2.14)$$

and the solution of (2.12) is given by

$$P(q,t) = N \exp[-U/2D] \sum_k b_k \exp[-\varepsilon_k t] \Psi_k(q) . \qquad (2.15)$$

As is well-known from molecular physics, $U(q)$ is very complicated and possesses a lot of minima. It will therefore take a long time until the lowest minimum $\varepsilon_o = 0$ is reached. Let us therefore consider the hopping between two eigenstates $\Psi_k \rightarrow \Psi_l$. The corresponding time is approximately [24]

$$\tau_{kl} = A \exp(\Delta U_{kl} / D) \qquad (2.16)$$

where ΔU_{kl} is the height of the threshold from state Ψ_k to Ψ_l. Therefore the time-dependent functional properties of the macromolecule depend explicitly on the shape of the potential U.

Because of the high efficiency of macromolecules we believe in an adaptation of these mechanical networks during evolution. This is connected with permanent alteration of the potential U with respect to the functional properties.

(ii) Reaction networks and ecological networks

The optimization of chemical networks played an important role in the evolution of life from the primal soup as well as in the development of the chemical mileu inside biological cells. The dynamics of ideal reactions is given by the balance equations

$$dx_i/dt = \sum_j a_{ij} x_j - \sum_{jk} b_{ijk} x_j x_k + c_i . \qquad (2.17)$$

It describes the gain or the loss by auto- or cross catalysis (a_{ij}), secondary collisions (b_{ijk}) or fluxes into the system ($c_i > 0$). Interpreting the concentration of the i-th component x_i as the state of the i-th node, (2.17) represents a special network dynamics. Under certain circumstances networks are capable of competition and selection, e.g. if the overall concentration is fixed, or some restrictions on the fluxes are included. A functional adaptation of the network or in other words, the occurrence of the fittest species x_i would take place by optimizing the selection rules for this species. Thus the minimization of the fluxes into the system or the maximization of the overall concentration yields drastic changes in the attractor landscape of the chemical network [4].

An important class of reaction networks are reaction diffusion systems. In this case, the index i labels the number of cells (nodes) into which the reaction vessel is divided and which are connected by diffusion fluxes. One-component systems are described by

$$dx_i/dt = -\partial_i U(\mathbf{x}) + \xi_i(t) = f (x_i) + \sum'_j D (x_j - x_i) + \xi_i(t) \qquad (2.18)$$

with

$$U(\mathbf{x}) = -\sum_i \int f(x_i) \, dx_i + \sum'_{ij} D/2 \, (x_j - x_i)^2 \qquad (2.19)$$

The dashed sums are to be extended only over the nearest neighbourhood and D is proportional to the Fickean diffusion constant. $f(x_i)$ stands for the local reaction rates of x_i in the i-th cell.

For bistable $f(x)$ with stable states $x^{(1)}$ and $x^{(3)}$ the number of stable states of the whole network depends strongly on the competition between reaction, diffusion and noise [28]. From a mean field approach it follows that if the noise intensity ε exceeds the critical value ε^{cr}

$$\varepsilon > \varepsilon^{cr} = D/8 \, (x^{(3)} - x^{(1)})^2 \qquad (2.20)$$

the system will be inhomogeneous , though every cell is bistable. For $\varepsilon <$ ε^{cr} the deepest minima are $x_1 = x_2 = ... = x_N = x^{(1)}$ and $x_1 = x_2 = ... = x_N = x^{(3)}$. The most frequent transition is then nucleation, i.e. the growth of adjacent cells into the new state by diffusion- induced transitions in the bistable cells [29].

The second interesting class of reaction- diffusion systems are excitable media. They possess a richer variety of possible attractors known from autowaves [17]. We will be dealing with excitable media in the following sections in terms of the possibilites they offer for information processing.

Let us mention that in many respects, ecological networks [5,7] show similar behaviour to chemical networks. The basic equation describing the dynamics of ecological communities is the so called Lotka-Volterra equation

$$dx_i/dt = x_i [a_i - \sum_j b_{ij} x_j] . \qquad (2.21)$$

In spite of the seemingly simple structure of the network (2.21) a deeper analysis shows a surprisingly rich structure [5,6] . A further step in evolution was the optimization of ecological communities with respect to the available space and the resources [3].

(iii) Electrical networks

Another class of networks of particular interest for technical applications are so-called RC-networks [30]. Applying Kirchhoff's famous laws, the dynamics of the voltage at the i-th node u_i is given by

$$du_i/ dt = I_i - u_i/ \tau_i + \sum_j J'_{ij} A(u_i) \qquad (2.22)$$

where $A(u)$ is a sigmoid function and J_{ij} symmetric connection matrix. This can be transformed to dynamic equations for the output voltages of the nodes

$$x_i = A(u_i)$$

resulting in

$$dx_i/dt = - \partial_i U(x) \tag{2.23}$$

where

$$U = \sum_i U_i(x_i) + \sum_{ij} J_{ij} x_i x_j. \tag{2.24}$$

Here $U(x)$ is a bistable potential with

$$du_i(x_i)/dt = [-I_i + A^{-1}(x_i) \ / \ \tau_i] / \langle dA/du \rangle \tag{2.25}$$

and $J_{ij} = J'_{ji}/\langle dA/du \rangle$ where $\langle dA/du \rangle$ is some average of the derivative. Including white noise in (2.23), we obtain the stationary solution of the corresponding Fokker-Planck equation

$$P^o(x) = N \exp(- [\sum_i U_i(x_i) + \sum_{ij} J_{ij} x_i x_j] / D). \tag{2.26}$$

The electrical network described above can solve complicated optimization problems as well as recognition problems if the "synaptic matrix" J is chosen in an appropriate way. Typically each element may have excitatory connections $J_{ij} > 0$ as well as many inhibitory connections $J_{ij} < 0$. In other words, the synaptic matrix reminds one in many respects of a spin glass coupling matrix. From this it follows that the potential of the problem $U(x)$ may have an extremely complicated (frustrated) character.

3. Pattern Processing with Excitable Media

In the following sections we want to present an algorithm for adapting an excitable medium to the task of pattern recognition. For this purpose we construct a learning machine working by means of a light-dependent excitable medium, which could be realized with a version of the Belousov-Zhabotinsky reaction (BZR) [31].

We shall show that reaction diffusion systems are good candidates for realizing learning machines. Real physical chemical systems are characterized by spatial flows of energy and matter in accordance with Fick's diffusion law. So the quick dynamics can be described in the most simple case with

$$\partial_t x = f(x, p) + D \Delta x . \tag{3.1}$$

Systems like this are networks of the type discussed in the section above as ideal reaction systems. Dividing the space into small boxes (nodes) we can model the reaction diffusion system (3.1) in terms of cellular automata. Conserving the essential properties of these coupled non-linear partial differential equations is a useful way to study them. A cellular automaton (CA) is a network of nodes with nearest neighbour interaction. They can thus be used to simulate reaction diffusion systems due to the local character of the diffusion. The dynamics of the state x_{tij} at the discrete time step t at the node at the i-th row and j-th column is given by

$$x_{t+1\ ij} = T\ (\ x_{tkl}\ ,\ p_{tij}\ ) \tag{3.2}$$

depending on the previous states of a set of surrounding nodes (x_{tkl}) and the local parameters p_{tij} [32]. On the one hand, one can model reality in spatio-discrete models such as CA, while on the other, CA could be constructed in such a way that it can perform natural processes for artificial information processing.

Non-equilibrium dissipative systems are capable of creating a rich variety of dynamic structures (cf. Fig. 2). One-component systems with quadratic characteristics show phase-waves switching the whole area from the unstable into the only stable state. Cubic characteristics make triggerfronts possible – these waves trigger the medium from the metastable into the stable state. This situation is due to nucleation phenomena and is currently being investigated intensively. Two-component systems show travelling pulses and auto-oscillations. They are characterized in Fig. 2-3 and 2-4 by their Null-clines, i.e. the diagram of the stationary homogeneous states ($f_i = 0$). The standard type is described with the dynamic equations

$$\partial_t\ x_1 = f_1 + D\ \Delta x_1 \tag{3.3}$$

$$\partial_t\ x_2 = \varepsilon\ f_2\ ;\qquad \varepsilon \ll 1\ .$$

The function $f_1 = 0$ is essentially of third order in x_1. In the Null-cline picture we divide it into three parts: stable inhibited (x_1 low), unstable (x_1 medium) and stable excited (x_1 high). In Fig. 2-3 $f_2 = 0$ crosses the stable inhibited part of $f_1 = 0$ – so this is the stable stationary state of the whole system. This state can be left for a short time t_+ to go into the excited state. If an excitation pulse runs through the medium, then spiral waves can be created too. These phenomena occur in pulse propagation in nerves and have been found to be responsible for heart activity. The case pictured in Fig. 2-4 is a visualization of the case where the stationary state is becoming unstable. So the system runs on a limit cycle – it is exhibiting auto-oscillations. Three-component systems are capable of chaotic behaviour. Some properties of hydrodynamic instabilities can also be described with such systems [33]. Basic features of reaction diffusion systems like (3.8) can already be studied in two-component systems – so-called excitable media [17].

For reaction diffusion systems Haken states the necessity of long-range interactions to realize tasks such as associative memory [34]. These long-range effects may be created by different time and length scales of different components in chemical reaction diffusion systems [35]. As Gerson and Mikhailov have suggested [36], this could be used in a sy-

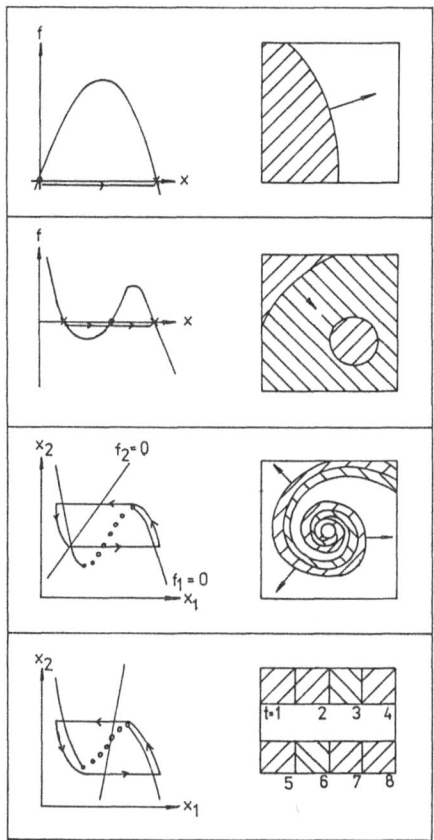

Fig. 2: Autowaves . One-component non-equilibrium reaction diffusion systems show phase and trigger fronts which can move of and stand still (Figs. 2-1 and 2-2). Two-component systems can form travelling pulses, of which spiral waves are a special case (Fig. 2-3), as are auto-oscillations (Fig. 2-4).

stem of chemical reagents for pattern recognition. Another basic feature of reaction diffusion systems is the influence of inhomogeneities. Kalnin [37] observed in experiments with the Benard effect several complicated self-reconstructing structures.

In recent papers [31, 38] the present authors have studied the influence of light on the light-dependent BZR. Here we shall show that light-dependent systems can be used for the construction of learning machines. Depending on the kind of inhomogeneous illumination, such systems show a rich variety of different attractors. The input into the learning machine is a light pattern projected onto the excitable medium. This serves as a boundary condition forcing the system to take on a distinct final state – the attractor, which is a dynamic structure. As output we treat these structures filtered by a set of threshold electrodes at marked points within the medium (Fig. 3). To obtain distinct input-

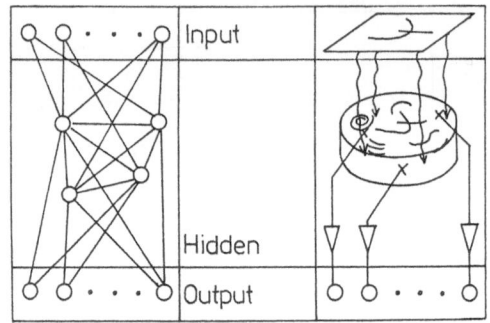

Fig. 3: Realization of learning machines. Left: a spin system - the whole set of spins is divided into input, hidden and output layer. Right: a reaction diffusion system, for instance an excitable medium is used as a learning machine: the coupling-in is realized by light; the hidden layer is a light-dependent excitable medium and the attractor is coupled out by a set of electrodes.

output relations from the learning machine we adapt the electrode positions [31]. During this slow dynamics a fitness function is maximized which is a measure for the distance between the outputs of different classes of input patterns in the filter space.

4. Automaton Model for Light-Dependent Excitable Media

The essential role in the concept of learning machines is played by dissipative dynamic effects taking place on a fast time scale, i.e. in the excitable medium, leading to a complicated attractor landscape. This section is devoted to such effects, to be studied in terms of a cellular automaton model.

Let us start by referring to experiments with the light-dependent BZR. It can be described with the extended Oregonator, a three-component model [31, 39]. The components are

$$x_1 \sim [\, H\,Br\,O_2\,]\; ;\quad x_2 \sim [\,Br^-\,]\; ;\quad x_3 \sim [\,Ru^{3+}\,] \tag{4.1}$$

and the corresponding dynamics (normalization according to Tyson [40])

$$\varepsilon\,\partial_t\,x_1 = x_2\,(q-x_1) + x_1\,(1-x_1) + D_1\,\Delta x_1 \tag{4.2}$$

$$\varepsilon'\,\partial_t\,x_2 = x_2\,(q+x_1) + f\,x_3 + I + D_2\,\Delta x_2$$

$$\partial_t\,x_3 = x_1 - x_3 + D_3\,\Delta x_3$$

with ε, ε', q, f and the light intensity I serving as space- and time-dependent parameters. The essential features of this system are bistability and parametric light-dependence. Making use of some scaling arguments ($\varepsilon \approx 4E\text{-}2$; $\varepsilon' \approx 2E\text{-}4$) and neglecting the diffusion of the slow component

x_3 we reduce (4.2) to the standard model for an excitable medium (3.3). In detail we state a second order rule for this network of coupled non-linear oscillators

$$h_{t+1\ ij} = h_{tij} + r^h\ (h^{ex}_{tij} - d^{sh}\ s_{tij} - h_{tij})\ , \tag{4.4}$$

$$s_{t+1\ ij} = sgn[\ s^{id}_{t+1\ ij} + j_{tij} - s^{cr}_{t+1\ ij}]\ .$$

First one calculates the value of the slow continuous component (the "field" h due to the component x_3) using a relaxation ansatz. There h^{ex} is a free parameter modelling the influence of light. The quick bistable component (the "spin" s = +1 or -1 due to x_1) immediately takes on the stable state due to the actual value of h. The term s^{id} models the hysteresis with the marginal stability points at $+h^{cr}$ or $-h^{cr}$. The only non-local operation is the calculation of the diffusion flow j by the sum of the differences of the eight nearest neighbours. s^{cr} refers to the unstable part of the Null-cline of s. The characteristics of the model are shown in Fig. 4. So we allow for hysteresis, diffusion-induced transitions and spatio-temporal light-dependence.

This algorithm is useful in studying the influence of light on the dynamics. First, we want to investigate the case of homogeneous illumination. For the remaining diffusion we obtain for our model the bifurcation diagram shown in Fig. 5. So we can switch the regime from excitation to inhibition via bistability or limit cycles varying the illumination intensity h^{ex}. Local node dynamics and global network dynamics for several regimes under homogeneous illumination are combined in Fig. 6. In all three cases the dynamics can be influenced by light. So we can manage the excitation time t_+ of a travelling pulse, the period $t_+ : t_-$ of an auto-oscillation or the propagation velocity v of a triggerfront.

In the following we consider the effects of inhomogeneity. In particular, we discuss two dissipative dynamic effects which are possible mechanisms for pattern processing: a damping and an amplification effect.

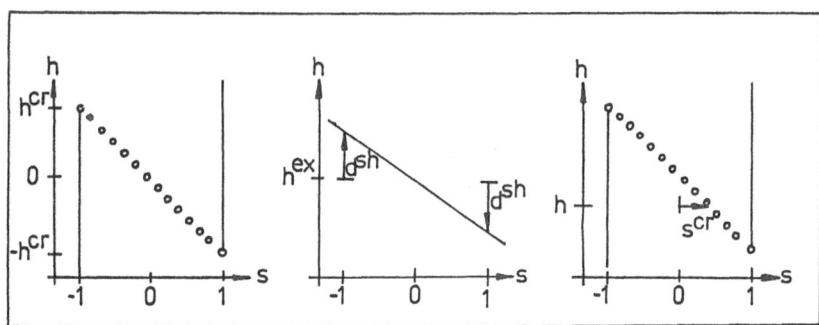

Fig. 4: Nullclines of the cellular automaton from sect. 4 . This representation visualizes the lines $f_i = 0$. Full lines mark stable while dotted lines indicates unstable parts. Left: the nullclines of the spin s; in the middle: nullclines of the field h. In the right diagram the threshold for diffusion-induced transitions s^{cr} is shown.

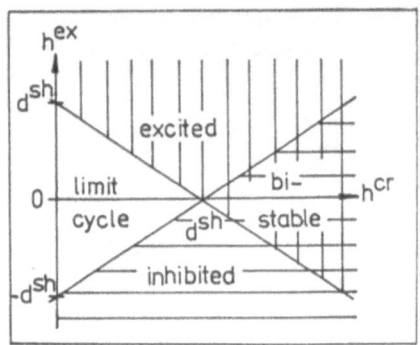

Fig. 5: Bifurcation diagram for the homogeneous case.

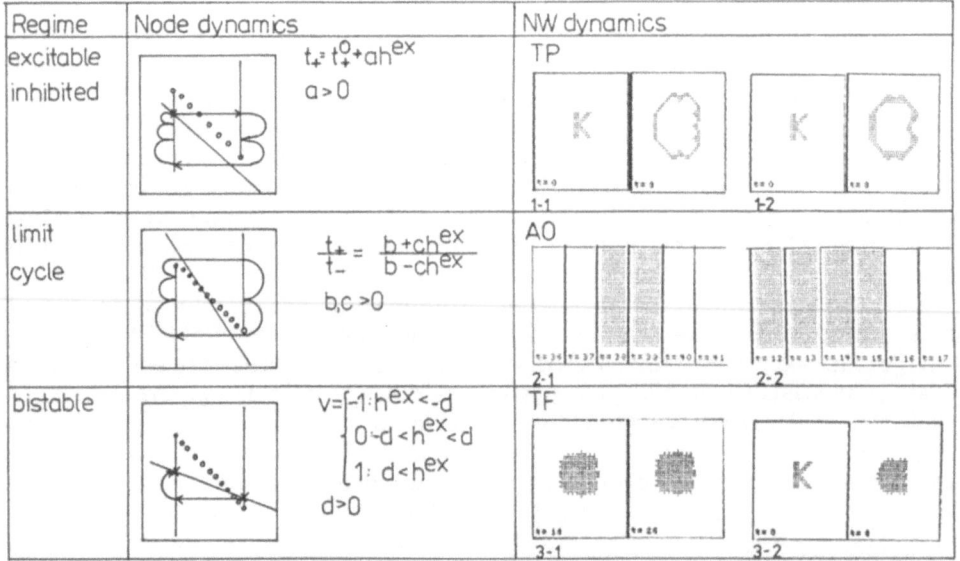

Regime	Node dynamics		NW dynamics
excitable inhibited		$t_+ = t_+^0 + ah^{ex}$ $a > 0$	TP 1-1 1-2
limit cycle		$\dfrac{t_+}{t_-} = \dfrac{b + ch^{ex}}{b - ch^{ex}}$ $b, c > 0$	AO 2-1 2-2
bistable		$v = \begin{cases} -1 : h^{ex} < -d \\ 0 : -d < h^{ex} < d \\ 1 : d < h^{ex} \end{cases}$ $d > 0$	TF 3-1 3-2

Fig. 6: Node dynamics and network dynamics Here we study the influence of light on the dynamics. In the excitable regime the excitation time t_+ increases with increasing illumination and so the travelling pulse broadens. The period of auto-oscillations can be changed - for $h^{ex} = 0$ the period is $t_+ : t_- = 2 : 2$, but for $h^{ex} > 0$ it is $4 : 2$. It is possible to force the triggerfront to stand still or to propagate; (further details in [38]).

Taking the dynamics of (4.4) in the continuous picture

$$\partial_t s = f + D \, \Delta s \tag{4.5}$$

$$\partial_t h = \varepsilon \, g$$

and letting, $\varepsilon \to 0$, we can eliminate h adiabatically and refer to the

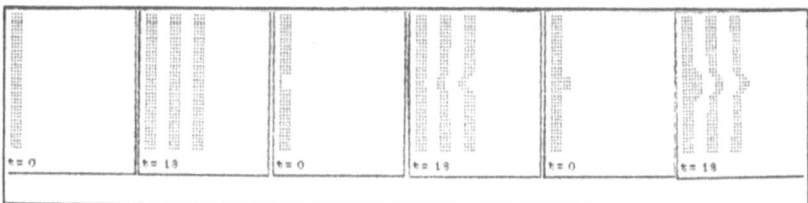

Fig. 7: Damping effect: Curved fronts become smoothed out.

results of Kawasaki et al. [41, 42] for the propagation velocity of a single triggerfront at the position

$$[\ x,y \ ] = [\ x, \ f(x,t) \ ]$$

$$\partial_t f \ / \ a = v_0 + \ D \ K \tag{4.6}$$

$$a = (\ 1 + (\partial_x f)^2 \ )^{1/2} \ ; \quad K = (\ \partial^2_x \ f \ ) \ / \ a^3$$

where v_0 is the wave front velocity and D the diffusion constant.

(i) Damping effect

According to (4.6) curved fronts become smoothed out. To demonstrate this effect we illuminate the medium with a pattern, thereby exciting it, while the surrounding medium remains in the excitable state. We thereby obtain a train of travelling pulses smoothing out their fronts (c.f. Fig. 7). This damping effect compresses several initial conditions into attractors represented by travelling trains of pulses.

(ii) Amplification effect

This effect separates initial conditions into different attractors if a threshold condition is fulfilled. Here we choose the illumination homogeneously. The initial condition is inhomogeneous – the pattern to be excited and the surrounding inhibited. At the end of the local excitation time t_+^{lo} a region of length Δu (the borderline of the pattern is parametrisized with u) is left inhibited. There one has

$$K(u) < K^{cr} \ (h) \ = \ - \ v_0 \ (h) \ / \ D \tag{4.7}$$

K (u) being the local curvature. If at the end of the refraction time of the local pattern t_-^{lo} some excited region is still present at the borderline, i.e. if

$$t_-^{lo} < \ t_{\Delta u/2}^{gl} \ + t_+^{gl} \tag{4.8}$$

the time $t_{\Delta u/2}^{gl}$ for crossing $\Delta u/2$ and the excitation time t_+^{gl} are large enough, a spiral wave can break into the local pattern (c.f. Fig. 8). Summing up: if the length of unexcited area is overcritical,

$$\Delta u > \Delta u^{cr}, \tag{4.9}$$

a pair of spiral waves is formed. Such dynamic structures can thus serve

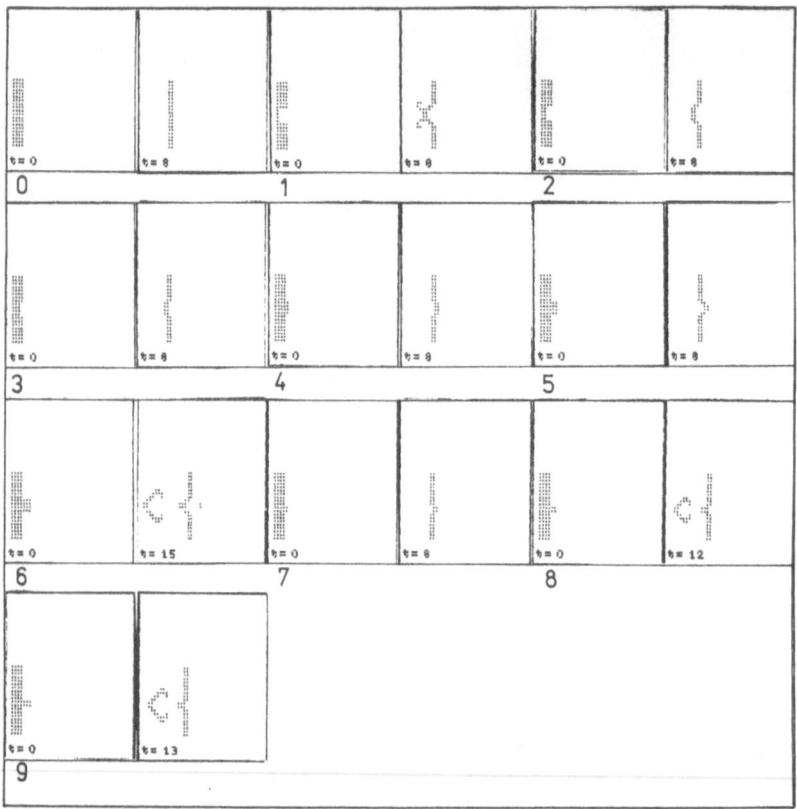

Fig. 8: Amplification effect. If the curvature of the initial condition is overcritical long enough for ($\Delta u > \Delta u^{cr}$) a pair of spiral waves arises, the length Δu of which is varied.

as good indicators for "free ends" of patterns because they amplify the property of overcritical curvature due to (4.7) [45].

5. Adaptation of the Chemical Network

Having discussed the possible realization of dissipative dynamic effects in excitable media, we want to construct and simulate a learning machine by means of double dynamics. The fast dynamics realizes sufficiently complicated input output relations depending on the system parameters. They are adapted in an overlayed slow process in agreement with the pattern processing task.

How is the learning machine to be organized? We construct the input $x^{(1)}$ as a strongly illuminated pattern within a moderately illuminated surroundings. This is due to the situation where the damping effect occurs. Consequently the dynamics results in periodic sequences of travelling pulses. So the hidden plane produces quite complicated structures.

The essential information is extracted by filtering this dynamic structure using a set of threshold electrodes. This procedure fulfills three tasks:

(1) Binary digits. The output of a threshold electrode at the position R_c is a binary digit

$$f_{tc} = H [s_{t R_c} - s^{cr}] \qquad (5.1)$$

where H is the Heavyside function. This is very important for the effective connection between experimental equipment and the computer.

(2) Dimensionality reduction. The spatio-temporal pattern is in general very complicated. If every cell can take on two states, the array has L^2 pixels and the period of the dynamic structure is T, then the number of different states for the dynamic medium is

$$n = 2^{L^2 T} / T . \qquad (5.2)$$

This enormous number (in our simulations the logarithmus dualis of n was $ld(n) \approx 8 \cdot 10^4$) is reduced by filtering through C threshold electrodes to

$$\tilde{n} = 2^{CT} / T \qquad (5.3)$$

(in the simulations ld ($\tilde{n}$) was 27.7).

(3) Selective projection. By varying the filter position R we can classify the patterns (c.f. Fig. 9).

In the operator picture the dynamics of the filter vector $f_t = (f_{tc})$ looks like

$$f_{t+1\ c} = \tilde{T} (x_{tkl} , p_{t R_c} , R_c) \qquad (5.4)$$

with the parameters

$$\tilde{p} = [p, R]; \qquad R = (R_c) . \qquad (5.5)$$

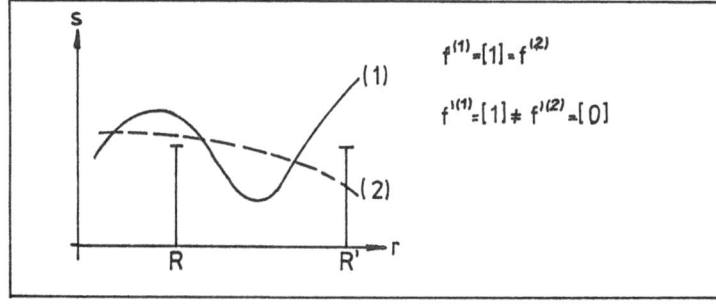

Fig. 9: Selective projection by threshold electrodes. For the filter position R the two concentration distributions $s^{(1)}$ and $s^{(2)}$ are mapped onto one point f in the filter space, but for the other position R', different points result. So different concentration profiles can be classified.

Thus we can fully determine the dynamics of the system with the extended parameter

$$\tilde{q} = [\ p,\ x_o\ ,\ x_{\partial A}\ ,\ R\ ] \tag{5.6}$$

including inner parameters p , initial and border conditions x_o and $x_{\partial A}$ and the filter position R.

For the damping effect we choose

$$\tilde{q} = [\ [\ x^{(1)},\ p'\ ],\ 0,\ \partial x = 0\ ,\ R\ ] \tag{5.7}$$

The corresponding regime of local excitation and global excitability we simulate with a special cellular automaton. This modification of an axiomatic model by Wiener and Rosenbluth (1946) [43] is represented in Fig. 10 as the state diagram. The resulting periodic pattern is mapped onto a matrix

$$F = (\ f_{t'}\ ) = (\ f_{t'\ c}\ )\ ; \qquad t' = t\ mod\ T \tag{5.8}$$

where t is the period.

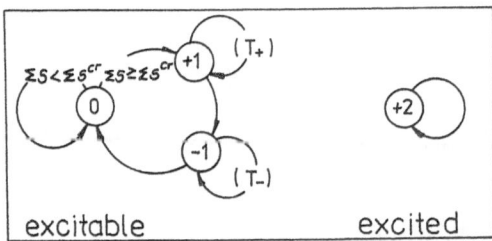

Fig. 10: State diagram of the cellular automaton from sect. 5. The input pattern remains in the excited state (s = +2). The surrounding is excitable; i.e. the resting state s = 0 can be left if the number of excited neighbours (s > 0) Σ s is overcritical ($\Sigma s > \Sigma s^{cr}$). In this case it stays t_+ steps in the excited state (s = +1) and t_- steps in the refractory state (s = − 1). The states of this cellular automaton are fully discrete ([s,c] $\in$ (−1,0,+1,+2)*(1,...,t_+ +t $_-$)).

In this manner we realize the input-output relations according to (3.1) (Fig. 11)

$$x^{(1)} \rightarrow F = x^{(o)}\ . \tag{5.9}$$

As a concrete task for the learning machine we want it to recognize the three letters "I", "O", and "X", each "disturbed" by 14% . How should the input-output relations be characterized in this respect? How should the adaptation procedure be formulated? What parameters should be adapted?

We decided to leave all parameters constant and to vary only the threshold electrode position R. This is because we are looking for the

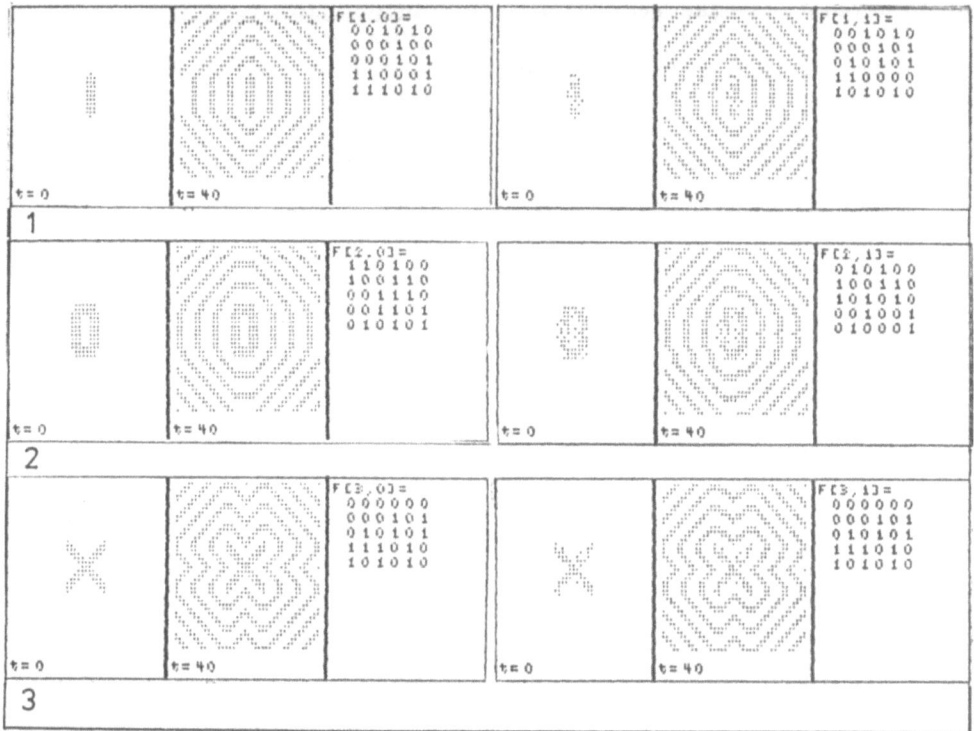

Fig. 11: Fast dynamics. The input is a constantly excited pattern - the three letters I, O and X. They are labeled i = 1, 2 and 3. In the chosen regime the excitable medium produces periodic patterns, which are mapped by a set of threshold electrodes onto a matrix $F[i, v]$ characterising the v-th variation of the i-th input.

attractors of the system of an excitable medium and threshold electrodes.

Treating the hidden element, the excitable medium, as a black box the input-output relations must be confirmed stastistically. If we want to classify some patterns, we give a set of inputs $x_{iv}^{(1)}$, which is the v-th variation of the i-th input, and obtain the corresponding output F_{iv} (Fig. 11). To measure the quality of the input-output relations we define a fitness function γ consisting of two parts

$$\gamma [R] = \gamma_1 + \gamma_2 \tag{5.10}$$

with

$$\gamma_1 = \left[\prod_{i_1 < i_2}^{I} \Delta (F_{i_1 0} , F_{i_2 0})^{2/I(I-1)} \right] / 2$$

$$\gamma_2 = - \sum_{i=1}^{I} \left[\sum_{v=1}^{V} \Delta (F_{i0} , F_{iv}) \right] / IV .$$

The first part γ_1 is the geometrically averaged distance between standard patterns. It is zero if different standard inputs ($v = 0$, $i_1 <> i_2$) give the same output ($F_{i_1 0} = F_{i_2 0}$). Here we have made use of the Hamming distance defined as

$$\Delta(\mathbf{G,H}) = \min_{t'} \left[\sum_{t=1}^{T} \sum_{c=1}^{C} \left(1 - \delta(g_{tc} , h_{t+t'} \, c) \right) \right] . \tag{5.11}$$

The second part γ_2 is the negatively counted arithmetically averaged distance between disturbed and standard outputs of one class (i fixed).

The number of ways of diostributing C threshold electrodes in the array of the excitable medium explodes exponentially with the number of threshold electrodes. It is

$$n_f = \binom{L^2}{C} . \tag{5.12}$$

So we treat this problem of complex optimization

$$\gamma \, [\, \mathbf{R} \,] \to \max. \tag{5.13}$$

with a mutation-selection strategy about which we do not want to expand in great detail.
A concrete run is shown in Fig. 12. After about 30 mutation steps and several selections we obtained fitness values in the interval

$$\gamma = 3.60 \; - \; 4.72 \; . \tag{5.14}$$

This result can be interpreted as follows (Fig. 13): the input patterns are well processed into several regions of the filter-space F.
These regions are well separated by empty space, because γ is a measure for the statistically averaged distance between different "output-clouds at the filter heaven". This means we have effectively pre-processed the task of the recognition of three patterns disturbed by 14%.

6. Conclusion

Here we have conceptualized and simulated a learning machine using an excitable medium for solving a special pattern recognition task. The architecture of the machine is derived from the double dynamics The fast dynamics is realized by a light-dependent excitable medium the state of which is filtered by a set of threshold electrodes at definite positions. Using the input pattern as strong illumination during the fast dynamics a dissipative dynamic effect damped the input perturbations. Optimizing the positions of the electrodes in a slow process we realized an effective pre-processing of the recognition problem.
We can therefore state:
(1) Using dissipative dynamic effects in networks with short-range interaction it is possible to pre-process the pattern in a proper way.

```
ADAPTION OF A REACTION-DIFFUSION-FILTER-SYSTEM FOR PATTERN RECOGNITION_____
   (Initial cond.=0, border cond.=0, inhom.=pattern)
Parameters:   Input (Patterns):     pat= 3;    pat_var= 5;    p_vary=5.8300E-01
   RDS (Exc.Med./Wiener-Rosenblueth):    t_exc= 2;    t_ref= 3;       T= 5;
      sig_cr=1.5000E+00;    T_max=40;
   Filters:     f=15;    f_comp= 6
   Adaption (Mutation/Selection):    f_mut= 5;    f_sel= 3;    f_best= 5
      f_vary=1.5000E+00;    ada_nr=44
   Output (Fitness/Mean Hamming distance (Reliability measure)):
      fit_max=10;
   Random number initialization:    init_rnd= 1100

Fitness dynamics:
   IfselnrIfmutnrI 0                      0.5                    1 Ifitbest    I
   I    I       I :   .   .   .    .   :   .   .    .    .   : I           I
   I   0I       I                                                 I           I
   I    I      0I  1 3 5    4                                      I2.8415E-01I
   I    I      1I     31     4 25                                  I3.3942E-01I
   I    I      2I   3  1     425                                   I3.3942E-01I
   I    I      3I        3    425                                  I3.3942E-01I
   I    I      4I        3     4 5                                 I3.5750E-01I
   I    I      5I        3    14 5                                 I3.5750E-01I
   I   1I       I                                                 I           I
   I    I      0I            54 31                                 I3.8613E-01I
   I    I      1I            54 31                                 I3.8613E-01I
   I    I      2I            54 31                                 I3.8613E-01I
   I    I      3I            54 31                                 I3.8613E-01I
   I    I      4I            54 312                                I4.1194E-01I
   I    I      5I            54 31 2                               I4.2494E-01I
   I   2I       I                                                 I           I
   I    I      0I                 425 1                            I4.7161E-01I
   I    I      1I                 425 1                            I4.7161E-01I
   I    I      2I                 425 1                            I4.7161E-01I
   I    I      3I                 425 1                            I4.7161E-01I
   I    I      4I                 425 1                            I4.7161E-01I
   I    I      5I                 425 1                            I4.7161E-01I
   I   3I       I                                                 I           I
   I    I      0I                  5 1                             I4.7161E-01I
   I    I      1I                  5 1                             I4.7161E-01I
   I    I      2I                · 5 1                             I4.7161E-01I
   I    I      3I                  5 1                             I4.7161E-01I
   I    I      4I                  5 1                             I4.7161E-01I
   I    I      5I                  5 1                             I4.7161E-01I
```

Fig. 12: Adaptation run. The head contains all parameters. Below the evolution of the fitnesses of the N_{best} best filters marked with their numbers is shown. Downwards the time in mutation-selection steps, right the value of the best fitness.

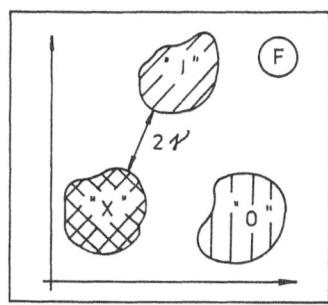

Fig. 13: Geometrical interpretation of the fitness

(2) The projection of the complicated spatio-temporal pattern by a set of threshold electrodes makes it possible to characterize it effectively.

(3) The adaptation of the input-output behaviour of a dynamic system is possible with the concept of double dynamics.

To complete the task of pattern recognition the several "clouds" in the F-space should be labelled using neighbourhood relations or decision planes. We want to emphasize that we have simplified this mathematical procedure by using the damping effect. The underlying cellular automaton realizing the excitable medium dynamics is easy to program and, because of its parallel character, very fast [44].

We should now like to outline some remaining problems and future prospects. The question of universality is an open one, but it should be possible to enlarge the complexity of the system - in general all parameters can be adapted

$$ w = \tilde{p} = [\ p, x_o, x_{\partial A}, R\ ] \tag{6.1} $$

to give rise to more complicated patterns, or more powerful dissipative dynamic effects. But the existence of a corresponding attractor landscape for a special information processing task is more or less a thesis - but nevertheless we want to underline the fact that light-dependent reaction diffusion systems are real physical chemical systems experimentally available and that they are adaptable by using the inhomogeneous illumination as a slow parameter.

The aim of this paper was to explain a possible mechanism of pattern processing in systems of excitable media and threshold electrodes. It is possible that a similar mechanism operates in the eye. The first stage of vision is characterized by light-dependent chemical excitable media [46] and a set of local threshold elements (rods and cones) playing the role of threshold electrodes [47]. In conclusion, it seems to be possible to manage pattern processing tasks in real physical chemical systems on the basis of the light-dependent BZR [39].

We should like to thank Prof. Kalnin (Leningrad) and Prof. Plath (Bremen) for their useful remarks.

References

1 H. Haken, Advanced Synergetics, Springer Series in Synergetics, **20**, Springer Verlag, Berlin, Heidelberg, New York (1983).

2 G. Nicolis, I. Prigogine, Self-Organization in Non-Equilibrium Systems, Wiley, New York (1977).

3 W. Ebeling, R. Feistel, Physik der Selbstorganisation und Evolution, Akademie-Verlag, Berlin (1982) and (1986).
R. Feistel, W. Ebeling, Evolution of Complex Systems, Verlag d. Wissenschaften, Berlin (1989).

4 M. Eigen, P. Schuster, The Hypercycle, Naturwiss. **64**, 541 (1977) and **65**, 341 (1978).
P. Schuster, Dynamics of Molecular Evolution, Physica **22D**, 100 (1986).

5 W. Ebeling, M. Peschel (edits.), Cooperation and Competition in Dynamical Systems, Akademie-Verlag, Berlin (1985).

6 M.Peschel, W.Mende, The Predator-Prey Model: Do we live in a Volterra World?, Akademie-Verlag, Berlin (1986).

7 W. Ebeling, Chaos, Ordnung und Information, Urania-Verlag, Leipzig (1986).

8 M.V. Wolkenstein, Entropie und Information, Nauka, Moskva (1986) (in Russ.), Akademie-Verlag, Berlin (1989), (German transl.).

9 H. Haken (Ed.), Complex Systems - Operational Approaches in Neurobiology, Physics and Computers, Springer Verlag, Berlin, Heidelberg, New York (1985).

10 H. Haken (Ed.), Computational Systems - Natural and Artificial, Springer-Verlag, Berlin, Heidelberg, New York (1987).
H. Haken (Ed.), Information and Self-Organization, Springer-Verlag, Berlin, Heidelberg (1988).
H. Haken, Der Computer erkennt mein Gesicht, Bild der Wiss. 8, 49 (1988), (in German).

11 J.S. Nicolis, Chaotic Dynamics Applied to Information Processing, Akademie-Verlag, Berlin (1987).

12 W.D. Hillis, The Connection Machine, Sci. American 256,(6), 86 (1987).

13 W. Ebeling, Physik komplexer nichtlinearer Netzwerke und neue Wege der Informationsverarbeitung, Sitzungsberichte der AdW der DDR, Akademie-Verlag, Berlin (1987).

14 T. Kohonen, An Introduction to Neural Computing, Neural Networks 1, 3 (1988).

15 S. Grossberg, Nonlinear Neural Networks, Neural Networks 1, 17 (1988).

16 J.J. Hopfield, Neural Networks and Physical Systems with Emergent Collective Computational Abilities, Proc. Natl. Acad. Sci. USA 78, 2554 (1982).
A. Engel. H. Englisch, A. Schütte, Improved Retrieval in Neural Networks with External Fields, Europhys. Lett. submitted.

17 Y.A. Vasilev, Yu.M. Romanovsky, D.S. Chernavsky and V.G. Yakhno, Autowave Processes in Kinetic Systems, Verl. d. Wiss., Berlin (1987).

18 M. Conrad, K.G. Kirby, Intraneuronal Dynamics as a Substrate for Evolutionary Learning, Physica 22D, 205 (1986).
M. Conrad, On Design Properties for Molecular Computers; Comm. of the ACM 28, 5 (1985).

19 F. Harary, R.Z. Norman, D. Cartwright, Structural Models, Wiley, New York (1985).

20 I. Sonntag, R. Feistel, W. Ebeling, Random Networks of Catalytic Biochemical Reactions, Biometr. J. 23, 91 (1981).

21 D.H. Ackley, G.E. Hinton, T.J. Sejnowski, A Learning Algorithm for Boltzmann Machines, Cogn. Sci. 9, 147 (1985).

22 W. Ebeling, Th. Boseniuk, Th. Pöschel, Simulation of Learning Networks, Proc. Parcella, Berlin (1988).

23 J.L. McClelland, D.E. Rumelhard (Eds.), Parallel Distributed Processing, MIT Press, Cambridge, Mass. (1986).

24 W. Ebeling, L. Schimansky-Geier, Transition Phenomena in Multi-Dimensional Systems - Models of Evolution, in: F. Moss, P. McClintock (Eds.), Noise in Nonlinear Dynamical Systems, Cambridge University Press, Cambridge, Mass. (1989).

25 J. Schnakenberg, Network Theory of Master Equation Systems, Rev. Mod. Phys. **48**, (1976).

26 W. Ebeling, I. Sonntag, Stochastic Description of Evolutionary Processes in Underoccupied Systems, BioSystems **19**, 91 (1986).

27 H. Risken, The Fokker-Planck Equation, Springer Series in Synergetics **18**, Springer-Verlag, Berlin, Heidelberg, New York (1984).

28 H. Malchow, L. Schimansky-Geier, Noise and Diffusion in Bistable Nonequilibrium Systems, Teubner-Texte zur Physik **5**, Teubner-Verlagsgesellschaft, Leipzig (1985).

29 L. Schimansky-Geier, W. Ebeling, Stochastic Theory of Nucleation in Nonequilibrium Bistable Reaction Systems, Ann. Physik, Leipzig **40**, 10 (1983). 30 J.J. Hopfield and D.W.Tank: Computing RC-Networks, Biol. Cybern. **52** , 141 (1985).

31 W. Ebeling, Pattern Processing and Optimization by Reaction Diffusion Systems, J. Stat. Phys. **45**, (5/6), 891 (1986).

32 Cellular Automata, Proc. of an Intern. Workshop, Los Alamos NM in Physica **10D**, (1984).

33 Y. Kuramoto, Chemical Oscillations, Waves, and Turbulence, Springer Series in Synergetics 19, Springer-Verlag, Berlin, Heidelberg, New York (1984).

34 H. Haken, Model of a Chemical Parallel Computer for Pattern Recognition and Associative Memory, J. de Chimie physique **84**, (11/12), 1289 (1989).

35 B.N. Belintzev, M.A. Livshitz, M.V. Volkenstein, Selforganization as a Result of Effective Nonlinearity, Phys. Lett. **82A**, (8), 20 (1981).

36 S.A. Gerson, A.S. Mikhailov, Possible Mechanism for the Analysis of patterns by Non-Equilibrium Biochemical Systems, Dokl. Akad. Nauk. SSSR **291**, (1) 228 (1986), (in Russ.).

37 A.A. Kalnin, On Possible Distributed Recording of the Signals in Non-Equilibrium Cooperative Systems (Structural Reflexia), Biofizika **29**, (1), 117 (1984).

38 Ch. Zülicke, L. Schimansky-Geier, Pattern Dynamics in Distributed Light-dependent Chemical Systems, Z. phys. Chemie, Leipzig, in press.

39 L. Kuhnert, L. Pohlmann, H.-J. Krug, G. Wessler, Repressing of Chemical Waves by Photochemical Inhibitor Releasing, in: W. Ebeling, H. Ulbricht (Eds.), Selforganization by Nonlinear Irreversible processes, Springer Series in Synergetics **33**, Springer-Verlag, Berlin, Heidelberg, New York (1986).

40 J.J. Tyson, A Quantitative Account of Oscillations, Bistability, and Traveling Waves in the BZR, in: M. Burger, R.J. Field (Eds.), Oscillations and Traveling Waves in Chemical Reactions, Wiley, New York (1985) p. 111.

41 K. Kawasaki, T. Ohta, Kinetic Drummhead Model of Interface (1), Theor.Phys. **67**, 147 (1982).

42 A. Engel, W. Ebeling, R. Feistel, L. Schimansky-Geier, Dynamics of Interfaces in Random Media, in: W. Ebeling, H. Ulbricht (Eds.), Selforganization by Nonlinear Irreversible Processes, Springer Series in Synergetics **33**, Springer-Verlag, Berlin, Heidelberg, New York (1986).

43 N. Wiener, A. Rosenblueth, The Mathematical Formulation of the Problem of Conduction of Impulses in a Network of Connected Exci-

table Elements, Specifically in Cardiac Muscle, Arch. Inst. Cadiol. Mex. **16**, 205 (1946).

44 T. Toffoli, M. Margolus, Cellular Automata Machines, MIT Press, Cambridge Mass, London England, (1987).

45 Ch. Zülicke, Formation of Spiral Waves from Inhomogeneous Initial Conditions, submitted to Z. Naturf.

46 N.A. Gorelova, and I. Bures, Spiral Waves of Spreading Depression in the Isolated Chicken Retina, J. Neurobiol. **6** (1983).

47 M.V. Volkenstein, Biophysics, Mir Publ. Moscow, (1983).

Synergetics Applied to Pattern Formation and Pattern Recognition

M. Bestehorn, R. Friedrich, A. Fuchs, H. Haken, A. Kuhn, and A. Wunderlin

Institut für Theoretische Physik und Synergetik, Universität Stuttgart,
Pfaffenwaldring 57/IV, D-7000 Stuttgart 80, Fed. Rep. of Germany

1. Introduction

Synergetics is the study and analysis of complex systems which are composed of many subsystems. These systems show the ability to selforganize spontaneously on macroscopic scales, i.e. we may observe highly ordered spatial, temporal, or spatio-temporal structures on scales which are much larger than the corresponding characteristic scales of the subsystems. As a result, we conclude that there must be very many subsystems coherently involved in order to produce these highly regular states on characteristic scales of the composed system. Synergetics offers a unified viewpoint to the emergence of such patterns and the underlying processes. Furthermore, powerful mathematical methods have been developed in this field to understand and predict the spontaneous occurrence of order, qualitatively as well as quantitatively.

The central aspects of the theory are introduced by the notions of instability, fluctuations, and slaving. Processes of self-organization in an open system always arise via an instability. Similarly, switching between self-organized states of a different macroscopic behaviour is again associated with instabilities. The system "notices" the presence of an instability through the fluctuations. Fluctuations may be considered as continually testing the stability of the present state of a system under externally given conditions. Eventually, slaving guarantees that the macroscopically evolving ordered state may be characterized by only a few collective variables, the so-called order parameters, which are generated by the system itself. It is understood that these order parameters enslave the subsystems to act in a coherent fashion and thus maintain the macroscopically ordered state under suitable external conditions. Furthermore they may be considered as a quantitative measure for the evolving order.

In parallel to its mathematical formulation, the slaving principle has been successfully applied to various systems treated in the natural sciences as well as to problems of the social sciences [1,2]. Laser action, for example, may be considered as a fundamental paradigm of slaving. The non-linear interaction between the laser-active atoms, which play the part of the subsystems here, and the light field generates a highly ordered state which is connected with the coherent laser action. Laser action is a particularly important example because its complete theory may be developed rigorously from the first principles of physics. In addition, we note that non-linear optical systems may be discussed along similar lines. Furthermore, the mathematical formulation of the principle has been applied to hydrodynamic systems, e.g. the Bénard and Taylor problem, solidification processes, flame instabilities, and atmospheric movements etc. There are applications in magnetohydrodynamics where the discussion of the generation of macroscopic stellar magnetic fields proves especially interesting. Self-organizing processes have also been discussed and identified in the field of plasma physics.

Applications, however, are not only restricted to physical systems. Indeed, the interdisciplinary aims are one of the most appealing aspects of synergetics. In chemistry, the Belousov-Zhabotinski reaction, the Briggs-Rauscher reaction and model reactions like the Brusselator and Oregonator have been analyzed using the slaving principle [1,2]. It has also been applied to biological problems, especially to problems of morphogenesis and is expected, as we shall discuss, to become an important tool in the analysis of brain functioning [3].

Finally, we mention applications in the social sciences: for example the problem of the formation of public opinion in sociology [4], and its use in the discussions of quantitative and qualitative models in economics [5].

Our article is organized as follows. First we shall introduce the mathematical formulation of the central principle of synergetics: the slaving principle. In view of the following applications, we restrict ourselves mainly to the treatment of deterministic systems. The slaving principle guarantees that we may apply methods of the qualitative theory of differential equations to describe macroscopically ordered structures observed in a large variety of systems showing the phenomenon of self-organization. We then apply these results to wave-like phenomena in synergetic systems which exhibit spatial symmetries. Specifically, we study waves in systems with SO(2)-symmetry and their occurrence in the Taylor problem and the baroclinic instability of large scale atmospheric eddies. Furthermore, we consider waves in systems exhibiting O(2)-symmetry and O(3)-symmetry. We then proceed to study travelling waves with cellular spatial structures in largely extended systems. An example of such systems is provided by the behaviour of binary fluid mixtures heated from below. Furthermore, we discuss - at least to some extent - an example of waves in biological systems, the so-called α-EEG waves, which are waves of the electric potential generated by brain activity. Finally, we present ideas of pattern recognition and a realization of an associative memory by a synergetic computer. These concepts have been developed quite recently.

2. The Basic Principle of Synergetics: The Slaving Principle

In the following we present an overview of the general mathematical methods of synergetics for systems where the details of the subsystems as well as their interactions are fully known. Full knowledge means that we know all the variables of the subsystems which we denote by q_i (i = 1, 2 ...) and the evolution of the single subsystems. If we put together all the variables into a state vector q in a state space Γ

$$q = (q_1, q_2, \ldots) , \tag{2.1}$$

its time evolution is typically governed by an equation of motion of the following general type

$$\dot{q} = N(q, \nabla, \{\sigma\}) + F . \tag{2.2}$$

Here N denotes a nonlinear vector field depending on q, which may additionally depend on spatial inhomogeneities symbolized through the ∇-symbol as well as a set of external parameters which have been abbreviated by $\{\sigma\}$. Symbolically, F summarizes the presence of fluctuations which are considered as a result of the action of the already suppressed microscopic degrees of freedom. We note that this typical form of the evolution equations can be generalized in various ways: We may include time-delay effects and non-local interactions in space, or we can formulate equations for the description of time discrete processes, etc.

We shall analyze the equations (2.2) in more detail. Generally, we are unable - at least by analytical methods - to describe the complete global behaviour of the solutions of (2.2) even if we confine ourselves to the deterministic part represented by N. We therefore concentrate on more simple subregions of the state space Γ which are of fundamental importance for an understanding of the underlying system: We consider stationary states. The most simple stationary state which we may expect as a solution describes a homogeneous situation in time and space. This state is usually referred to as the thermodynamic solution branch of the system. However, the method can be generalized considerably, i.e. we may even consider stationary states which are not homogeneous in space or in time (periodic and quasi-periodic states). For the present however, we confine ourselves to the most simple stationary state which is homogeneous in time and space. We construct this state as a solution of

$$\dot{q} = 0 \quad \text{--->} \quad q = q_o . \tag{2.3}$$

We now consider the stability of the state (2.3). When we change the external conditions by varying the control parameters, the stationary state q_o may lose its stability. Mathematically, this means that we are looking for instability regions in the parameter space spanned by $\{\sigma\}$. A general method of achieving this is provided by linear stability analysis. We are looking for the behaviour of small deviations $\underset{\sim}{v}$ from the stationary state q_o, i.e.

$$q = q_o + \underset{\sim}{v} . \tag{2.4}$$

Then we immediately find the equations of motion for $\underset{\sim}{v}$ from (2.2):

$$\dot{\underset{\sim}{v}} = L\underset{\sim}{v} + O(\|\underset{\sim}{v}\|^2) , \tag{2.5}$$

where, for simplicity, we have not taken the fluctuations into account. The elements of the linear matrix L are given by

$$L = (L_{ik}) \quad \text{and} \quad L_{ik} = \partial N_i/\partial q_k|_{q=q_o} . \tag{2.6}$$

Obviously, the matrix L is still a function of the stationary state q_o and may depend on gradients, and is a function of the external parameters $\{\sigma\}$. There exists a well-known algebraic procedure to solve (2.5). The ansatz

$$\underset{\sim}{v} = \underset{\sim}{v}_o \exp(\lambda t) \tag{2.7}$$

transforms (2.5) into a linear algebraic eigenvalue problem from which we obtain the set of eigenvalues λ_i and the corresponding eigenvectors, which we denote by $\underset{\sim}{\Phi}_i$. We assume that they form a complete set. Instabilities in the space of the external parameters are then indicated by the condition

$$\text{Re} \{\lambda_i\} = 0 . \tag{2.8}$$

We summarize the information obtained from our linear stability analysis. Firstly, we find the regions of instability in the space of the external parameters. Secondly, we find the eigenvectors which we interpret as the collective modes of the system and finally, we obtain locally the directions in Γ along which the stationary state may become unstable.

The next step of the systematic treatment consists of a complete non-linear analysis of (2.2). We note that the solutions of the full nonlinear equations can be written in the form

$$q = q_o + \underset{\sim}{v} \quad \text{and} \quad \underset{\sim}{v} = \sum_i \xi_i(t) \underset{\sim}{\Phi}_i . \tag{2.9}$$

We therefore transform the original equation (2.2) into an equation for the amplitudes of the collective modes $\xi_i(t)$:

$$\xi_i(t) = \Lambda_{ik}\xi_k(t) + H_i(\{\xi_i\}) + \bar{F}_i . \tag{2.10}$$

Here Λ is a diagonal matrix, or is at least of the Jordan canonical form, H_i

166

transformed fluctuating forces. We note that the set of equations (2.10) is still exact.

We thus arrive at the central step, which results in the formulation and application of the slaving principle. As mentioned previously, slaving means that in the vicinity of a critical region in the space of external parameters only a few modes, the so-called order parameters, dominate the behaviour of the complex system on macroscopic scales. Obviously, the linear stability analysis already indicates which amplitudes of collective modes will finally become the order parameters. Indeed, the linear movement yields a separation in the time scales: The modes which become unstable will move on a very slow time scale because of $|\text{Re}\lambda_u| \sim 1/\tau_u$, where we have identified the index i with u to exhibit unstable behaviour. In the vicinity of the critical point, τ_u becomes extremely large. On the other hand, the modes still remaining stable have a comparably short time scale which we denote by τ_s. Therefore in the instability region we find a hierarchy in the time scales

$$\tau_u \gg \tau_s \quad . \tag{2.11}$$

Taking into account the interaction of stable and unstable modes, we expect that in the long term behaviour of the system the stable variables behave in a way which is completely determined by the unstable modes; the order parameters. This observation may be cast into mathematical terms as we now demonstrate. We may split the set of $\{\xi_i\}$ into a set of modes which will become unstable and is denoted by the vector $\underset{\sim}{u}$ and the remaining set of stable modes which we collect into the vector $\underset{\sim}{s}$. Accordingly, we split the set of equations (2.10) into

$$\dot{\underset{\sim}{u}} = \Lambda_u \underset{\sim}{u} + \underset{\sim}{Q}(\underset{\sim}{u},\underset{\sim}{s}) + \underset{\sim}{F}_u \tag{2.12}$$

$$\dot{\underset{\sim}{s}} = \Lambda_s \underset{\sim}{s} + \underset{\sim}{P}(\underset{\sim}{u},\underset{\sim}{s}) + \underset{\sim}{F}_s \quad . \tag{2.13}$$

Slaving of the stable modes through the order parameters now means

$$\underset{\sim}{s} = \underset{\sim}{s}(\underset{\sim}{u},t) \quad . \tag{2.14}$$

Equation (2.14) expresses the fact that the values of the stable modes are completely determined by the instantaneous values of the order parameters u. With the result (2.14), the complete set of equations (2.12) and (2.13) can be drastically simplified. In fact we can use this result to eliminate the stable modes from the equations for the order parameters to arrive at the equation

$$\dot{\underset{\sim}{u}} = \Lambda_u \underset{\sim}{u} + \underset{\sim}{Q}(\underset{\sim}{u},\underset{\sim}{s}(\underset{\sim}{u},t)) + \underset{\sim}{F}_u \quad , \tag{2.15}$$

which is the final order parameter equation. We observe that the idea of slaving generally leads to a drastic reduction in the degrees of freedom of the system which, however, now yields a complete description of the system on macroscopic scales. The next step, then, consists in the solution of the order parameter equation (2.15) and in the identification of the macroscopically evolving ordered states which correspond to definite patterns or the functioning of the system. The central problem which still remains to be solved resides in the construction of (2.14).

In the construction of (2.14), our interest is devoted to the long term behaviour of the slaved modes. We can formally integrate (2.13) in the following way:

$$\underset{\sim}{s} = \int_{-\infty}^{t} \exp[\Lambda_s(t-\tau)]\{\underset{\sim}{P}(\underset{\sim}{u},\underset{\sim}{s}) + \underset{\sim}{F}_s\}_\tau \, d\tau \quad . \tag{2.16}$$

By means of (2.16), we define the operator $(d/dt - \Lambda_s)^{-1}$:

$$\underline{s} = (d/dt - \Lambda_s)^{-1} [\underline{P}(\underline{u},\underline{s}) + \underline{F}_s] \quad . \tag{2.17}$$

We mainly consider the purely deterministic case where F_s vanishes (generalizations are given in [2,6,7]). We further introduce the operator

$$(d/dt) = \underline{Q}\partial/\partial\underline{u} \tag{2.18}$$

and in addition

$$(d/dt - \Lambda_s)^{-1}_{(o)}\underline{P} = \int_{-\infty}^{t} \exp[\Lambda_s(t-\tau)] \ \underline{P}(\underline{u},\underline{s})_t d\tau \quad . \tag{2.19}$$

A partial integration in (3.2) can now be performed using the above defined operators in the following way:

$$(d/dt - \Lambda_s)^{-1} = (d/d\underline{t} - \Lambda_s)^{-1}_{(o)}$$

$$- (d/dt - \Lambda_s)^{-1} (d/dt) (d/dt - \Lambda_s)^{-1}_{(o)} \quad . \tag{2.20}$$

We now may use (2.20) to construct a systematic iteration scheme to find (2.14). We take the ansatz

$$\underline{s} = \sum_{n=2}^{N} \underline{c}^{(n)} \quad , \tag{2.21}$$

where $\underline{c}^{(n)}$ is precisely of order n in $\underline{u}$. Correspondingly, we define $\underline{p}^{(n)}$ and $(d/dt)^{(n)}$. Inserting (2.20) and (2.21) into (2.16) we obtain the solution in the form

$$\underline{c}^{(n)} = \sum_{m=0}^{n-2} (d/dt - \Lambda_s)^{-1}_{(m)} \ \underline{p}^{(n-m)} \tag{2.22}$$

where we have defined

$$(d/dt - \Lambda_s)^{-1}_{(m)} = (d/dt - \Lambda_s)^{-1}_{(o)} \sum_i \prod_i [(-d/dt)^{(i)} (d/dt - \Lambda_s)^{-1}_{(o)}] \tag{2.23}$$

and the product over i has to be taken in such a way that $i \geq 1$ and $\sum i = m$. The sum means summing over all different products. Equation (2.23) provides us with a systematic method of constructing (2.14) and it has been generalized in various ways, especially by taking into account the fluctuations [2,7]. Furthermore, we need not restrict ourselves to the stationary time-independent states q_o. Finally, we may also take into account, in largely extended systems, slow spatial variations by including the effects of finite bandwidths [8].

Here, we have presented a mathematical method introduced by the idea of slaving and order parameters. We observe a drastic reduction in the degrees of freedom by restricting ourselves to the order parameters which are responsible for the macroscopically evolving structures. We would like to mention that our procedure can be generalized in various ways. For example we may take full account of fluctuations. In the following we apply these ideas to wave-like phenomena in synergetic systems.

3. Instabilities and Symmetries

It is obvious that the various forms of spatial patterns arising as a result of spontaneous self-organization are also related to the geometry of a system. Thus, symmetry considerations play an important role in the investigations of the order parameter equations describing a nonequilibrium phase transition. The symmetries of a system manifest themselves in the structure of the order parameter equations. It is possible to derive normal forms of the order parameter equations, characterizing instabilities in the presence of symmetry. This may be done in a similar way to the investigation of instabilities in the presence of symmetry, using static bifurcation theory, see [9]. To this end, we briefly describe the main steps. We consider an evolution equation of the form

$$d_t \underline{v} = L \underline{v} + \underline{N}[\underline{v}, \sigma] \tag{3.1}$$

of a system which is invariant with respect to transformations of a certain symmetry group G with elements g. In the state space of the system, this group induces the representation D_g. The system is invariant with respect to the symmetry transformation g and, as a consequence, the evolution equation exhibits the invariance

$$d_t D_g \underline{v} = L D_g \underline{v} + \underline{N}[D_g \underline{v}, \sigma] \tag{3.2}$$

and we have the two conditions

a) $\quad D_g L = L D_g$, $\qquad$ b) $\quad \underline{N}[D_g \underline{v}, \sigma] = D_g \underline{N}[\underline{v}, \sigma]$. $\tag{3.3}$

From condition a) it follows that the normal modes Φ_j of the linear operator L belong to irreducible representations of the group G. The state space of the system is decomposed into a direct sum of sub-spaces, spanned by groups of normal modes transforming irreducibly under the group transformations. Close to instability, the state vector $\underline{v}$ can be represented, according to the slaving principle, in the form

$$\underline{v} = \sum_u \xi_u \Phi_u + \sum_s \xi_s(\xi_u) \Phi_s \tag{3.4}$$

As shown above, the amplitudes of the unstable modes obey the order parameter equations

$$d_t \xi_u = \lambda_u \xi_u + h_u(\xi_u) \tag{3.5}$$

where $h_u(\xi_u)$ contains the nonlinearities. The operator D_g, corresponding to the symmetry operation g, transforms a solution $\underline{v}$ to another solution $D_g \underline{v}$ which may be obtained from (3.4) in the form

$$D_g \underline{v} = \sum_u \xi'_u \Phi_u + \sum_s \xi'_s(\xi'_u) \Phi_s \tag{3.6}$$

by the relations

$$\xi'_u = \sum_{u_1} D^{(u)}_{uu_1} \xi_{u_1} , \qquad \xi'_s = \sum_{s_1} D^{(s)}_{ss_1} \xi_{s_1}$$

$$D^{(u)}_{uu_1} = <\Phi^+_u|D_g \Phi_{u_1}>, \quad D^{(s)}_{ss_1} = <\Phi^+_s|D_g \Phi_{s_1}> \quad . \tag{3.7}$$

The matrices $D^{(s)}_{ss_1}$, $D^{(s)}_{ss_1}$ are representations of the group G in the sub-spaces spanned by the unstable and stable modes of the linear operator L. Due to the invariance of the evolution equation (3.1) the order parameter equation has to be invariant with respect to the transformations (3.7):

$$\sum_{u_1} D^{(u)}_{uu_1} h_{u_1}(\xi_u) = h_u(\sum_{u_1} D^{(u)}_{uu_1} \xi_{u_1}) \quad . \tag{3.8}$$

The validity of this relation may also be proved by an explicit construction of the nonlinearities along the lines indicated in Section 2. The invariance condition allows one to establish the general form of the nonlinearities of the order parameter equations for systems which are invariant under symmetry operations. Usually, the nonlinearities are approximated by polynomials so that the above invariance conditions lead to relations among the coefficients of these polynomials. The order parameter equations are then reduced to normal forms, allowing a discussion of instabilities in quite different systems exhibiting the same symmetries. From this point of view, we discuss in the following, instabilities resulting in various wave-like structures in systems characterized merely by their symmetries.

4. Waves in Systems with SO(2)-Symmetry

First, we consider the oscillatory instability of a state of a synergetic system with SO(2)-symmetry. This state, as well as the evolution equation for its deviation $\underline{v}(\underline{r},t)$, has to be invariant with respect to the arbitrary transformation $\phi \rightarrow \phi + \phi_0$ of the cyclic variable ϕ. The normal modes corresponding to the linear eigenvalue problem arising in the examination of the stability of the basic state then have, as a result of the symmetry, the form

$$\Phi_{jm}(\underline{r}) = \Phi_{jm}(r,z) \exp(im\phi) \quad m=0,\pm1,\pm2\ldots . \tag{4.1}$$

We are concerned with an instability of modes with $m\neq0$, i. e. with an instability breaking the SO(2)-symmetry. The corresponding eigenvalue λ_{jm} is assumed to be complex in order to obtain an oscillatory instability. The mode $\Phi_{j,-m}(\underline{r})$ is a normal mode to the conjugate complex eigenvalue. Therefore, we have an instability induced by one pair of modes. Due to the slaving principle, the state vector close to instability is given by

$$\underline{q}(r,\phi,z,t) = \underline{q}^0(r,z,t) + \xi_u(t)\Phi_{um_u}(r,z) \exp(im_u\phi) \tag{4.2}$$

$$+ \sum_{s,m_s} \xi_s[\xi_u]\Phi_{sm_s}(r,z) \exp(im_s\phi) + c.c .$$

A transformation $\phi \rightarrow \phi + \phi_0$ changes the state vector $\underline{q}(r,\phi,z,t)$ to $\underline{q}(r,\phi+\phi_0,z,t)$. Due to symmetry, this is also a possible state of the system and therefore obeys the same evolution equation. The state vector $\underline{q}(r,\phi+\phi_0,z,t)$ can also be generated according to (3.7) by the transformation

$$\xi_{jm} \rightarrow \xi_{jm} \exp(im\phi_0) . \tag{4.3}$$

The order parameter equation has to be invariant under the transformations $\xi_u(t)\rightarrow\xi_u(t)\exp(im_u\phi_0)$. This invariance condition allows one to establish the general structure of the order parameter equation:

$$d_t \xi_u(t) = (\varepsilon+i\omega_0) \xi_u(t) - \xi_u(t) h(|\xi_u(t)|^2) . \tag{4.4}$$

Here, h is an arbitrary function of $|\xi_u(t)|^2$. This function may be approximated by $a|\xi_u(t)|^2$ close to instability, if the real part of the coefficient a is larger than zero. After transients have died away, the solution of this equation has the form

$$\xi_u = r \exp\{i\omega t\}$$

$$r = [\varepsilon/\mathrm{Re}(a)]^{1/2} , \quad \lambda = \lambda_0 - \mathrm{Im}(a) r^2 . \tag{4.5}$$

It can also be shown as a consequence of symmetry that ξ_s=const $\exp(im_s/m_u \omega t)$. The state vector takes the form of a rotating wave:

$$\underline{q}(r,\phi,z,t) = \underline{q}(r,\phi+\omega/m_u\, t,z) \tag{4.6}$$

In a co-rotating coordinate system the state vector is stationary.

For the more complicated case where an instability is induced by several pairs of unstable modes with different values of m, the general form of the order parameter equations can easily be established by exploiting the condition of invariance of the equations under the transformations (3.7). The non-linearities in these equations can again be approximated by polynomials, which contain only a few coefficients. One can classify the solutions of the order parameter equations according to these coefficients and obtain a classification of the possible kinds of behaviour which are expected to occur close to such an instability. For the solution of a special problem, the coefficients of the Taylor expansion of the nonlinearities have to be determined starting from the basic evolution equations. We consider two examples in more detail.

The first example we discuss is that of waves arising in the Taylor-Couette experiment of hydrodynamics [10]. In this experiment, a homogeneous viscous fluid is contained between two concentric cylinders where the inner cylinder rotates with a certain angular velocity measured, in dimensionless units, by the Reynolds number $Re = d\Omega r/\nu$ (d is the width of the gap between the cylinders, Ω the angular velocity and r the radius of the inner cylinder, ν is the kinematic viscosity of the fluid). For small Reynolds numbers, one observes the fluid flowing in purely azimuthal direction (Couette flow). Above a certain value, a transition occurs to the Taylor vortex flow. This flow consists of axisymmetric toroidal vortices. It is stationary and possesses, in the idealized case of infinitely long cylinders, rotational symmetry in the azimuthal direction. By increasing the Reynolds number, this SO(2)-symmetric state undergoes an oscillatory instability in the so-called wavy vortex flow. A non-axisymmetric disturbance characterized by a certain azimuthal wave number m, is superimposed on the stationary Taylor vortices travelling with a certain wave speed in the azimuthal direction. From the above discussion of the oscillatory instability, it directly follows that the wavy vortex flow of the Taylor-Couette experiment is a rotating wave. Thus the order parameter equation takes the form given above. The detailed calculations have to determine the values of the critical Reynolds number for the onset of instability, the azimuthal wave number m of the critical mode, its linear growth rate ε and its wave speed λ_0 as well as the coefficient a. The calculations proceed as follows [11]: The evolution of the flow is described by the Navier-Stokes equations for the velocity field $\underline{v}(\underline{r},t)$ and the pressure $p(\underline{r},t)$ for the case of an incompressible fluid:

$$\partial_t\,\underline{v}(\underline{r},t) + (\underline{v}(\underline{r},t)\nabla)\,\underline{v}(\underline{r},t) = \nu\,\Delta\,\underline{v}(\underline{r},t) - \nabla\,p(\underline{r},t)$$

$$\tag{4.7}$$

$$\mathrm{div}\,\underline{v}(\underline{r},t) = 0 \quad .$$

Since it is well-known from experiments that the wave number in the axial direction of the Taylor vortex flow, as well as of the wavy vortices, remains fixed in the range of Reynolds numbers under consideration, the flow is assumed to be strictly periodic in z-direction. In a first step the velocity field $\underline{U}(r,z)$ of the axisymmetric Taylor vortex flow is determined. This has to be done numerically. In the second step, the stability of this flow is investigated with respect to nonaxisymmetric disturbances with several azimuthal wave numbers m. The threshold of the instability is determined by the condition that the real part of the eigenvalue λ vanishes for the most unstable modes. This defines the critical Reynolds number as well as the wave speed ω_0 and the linear growth rate of the critical modes. Results of the calculations are shown in Fig. 1 for the case of a radius ratio of the two cylinder radii of .8. The critical modes possess the azimuthal wave number m=1. In the third step, the stable modes are eliminated adiabatically and the coefficient a is determined. The solution of the order parameter equation then determines the amplitude and the wave speed of the wavy vortex flow close to threshold. The spatial structure of the rotating wave is exhibited in Fig. 2.

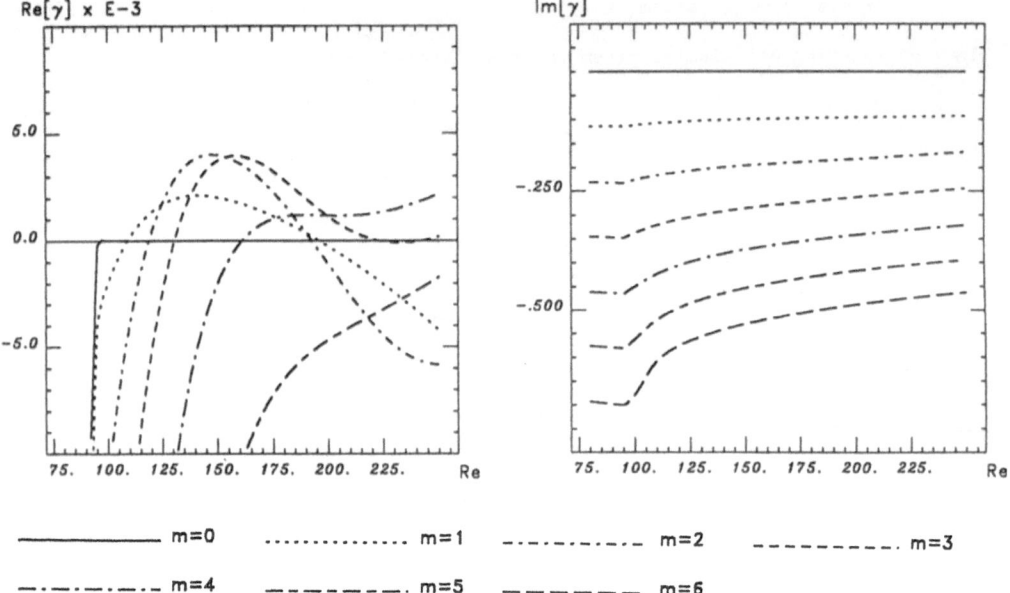

Fig. 1: Real and imaginary parts of the eigenvalues determining stability
of Taylor vortex flow for disturbances with azimuthal wave numbers
m=0,1,...,6; after [11]

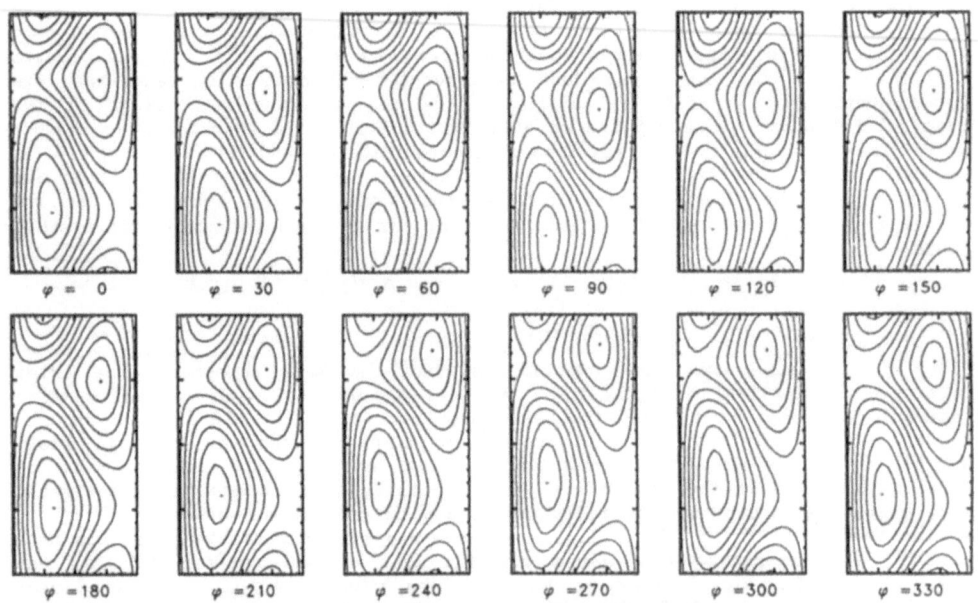

Fig. 2: Contour plots of the azimuthal component of the velocity field
for Wavy Vortex Flow with azimuthal wave number m=1 at different
angles ϕ (Reynolds number Re=150, η=.8); after [11]

In our second example, we discuss a model used for the explanation of the formation of the large scale motions of the earth's atmosphere. The atmospheric eddy structure is the dominating element of weather activity. The most relevant instability responsible for the formation of coherent atmospheric patterns is evidently the baroclinic instability. It occurs in rotating, vertically stably stratified fluids under the influence of strong Coriolis forces and a horizontal temperature gradient. In the atmosphere the horizontal temperature gradient is caused by latitude-dependent heating by the sun. A weak horizontal temperature gradient results in a pressure gradient between the equator and the poles. This pressure gradient is balanced by the Coriolis forces caused by an almost zonal wind sheared in the vertical direction. This zonal wind is subjected to the baroclinic instability if the temperature gradient exceeds a critical value of approximately 6 K/1000 km.

The following results have been obtained on the basis of the quasi-geo-strophic two-layer model, originally introduced by Phillips [12]. The large scale eddies generated by the baroclinic instability occur in the middle latitudes between 30^0-60^0. The two-layer model approximates this geometry by a rotating rectangular channel. Let us consider two stably stratified layers of immiscible fluids of different densities. In the limit of fast rotation, the pressure amplitudes in the two fluid layers ψ_1, ψ_2 obey the quasi-geostrophic equations

$$(\frac{\partial}{\partial t} + \frac{\partial \psi_1}{\partial x}\frac{\partial}{\partial y} - \frac{\partial \psi_1}{\partial y}\frac{\partial}{\partial x}) \; [\; \Delta\psi_1 + F \; (\psi_2 - \psi_1) + \beta y] + r \; \Delta \; \psi_1 = 0 \quad ,$$

$$(\frac{\partial}{\partial t} + \frac{\partial \psi_2}{\partial x}\frac{\partial}{\partial y} - \frac{\partial \psi_2}{\partial y}\frac{\partial}{\partial x}) \; [\; \Delta\psi_2 + F \; (\psi_1 - \psi_2) + \beta y] + r \; \Delta \; \psi_2 = 0 \quad .$$

$$(4.8)$$

F is the Froude number, which is a dimensionless measure of the rotation rate. The parameter β is a measure of the variation of the Coriolis parameter in the north-south direction (y-direction), which takes into account the variation of the Coriolis force due to the curvature of the earth. The parameter r describes dissipative effects like the transfer of energy to small-scale turbulent motions. The pressure fields have to satisfy periodic boundary conditions in x-direction:

$$\psi_i(x,t) = \psi_i(x+L,t) \quad . \tag{4.9}$$

Here, L is the length of the channel with normalized length d=1 in the north-south direction (y-direction). Additionally, the side walls of the channel (y=0, y=1) are assumed to be impenetrable. These conditions lead to the boundary conditions

$$\frac{\partial \psi_1}{\partial x} = \frac{\partial \psi_2}{\partial x} = 0 \text{ for } y=0, \; y=1 \tag{4.10}$$

valid in the geostrophic approximation. In the quasi-geostrophic approximation, different boundary conditions have to be used [13]. Due to the periodicity in the x-direction, the system exhibits SO(2)-symmetry.

The basic flow described by the two-layer model is given by

$$\psi_1 = -Uy \quad , \quad \psi_2 = Uy \tag{4.11}$$

where U is constant. This flow consists of a zonal shear flow; the wind flows in opposite directions in the two layers. In the stable situation, the resulting Coriolis forces just balance the pressure gradient in the y-direction. If this shear flow exceeds a critical value U_{crit}, baroclinic instability sets in. The linear stability analysis of this shear flow leads to normal modes of the form

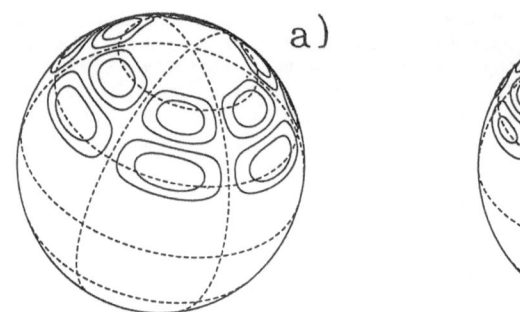 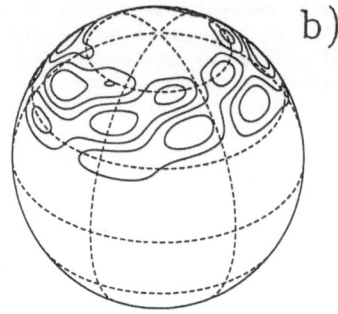

Fig. 3: a) An example of a basic flow pattern of the baroclinic instability in middle latitudes (meridional wave number m=2, zonal wave number n=3).
b) Typical flow pattern;
after [13]

$$\Phi_{mn}(x,y) = N \sin(m\pi y) \exp(i\, n/L\, x) \quad \text{for } n\neq 0$$

$$\Phi_{mn}(x,y) = N \cos(m\pi y) \qquad\qquad \text{for } n=0 \qquad\qquad (4.12)$$

$$m = 1,2,3\ldots \quad ; \; n = \pm 1,\pm 2, \ldots \quad .$$

Each mode $\Phi_{mn}(x,y)$ describes an eddy pattern with n pairs of high and low pressure centers in the zonal direction and m pressure centers in the meridional direction, see Fig. 3. Usually, one pair of modes m,±n becomes unstable for a certain strength of the basic shear flow. By varying the additional parameters F, r/β, and L, it is possible to destabilize several modes simultaneously. For example, it is possible to find certain sets of parameters, where two, three, and four pairs of modes become unstable. We consider the case of four pairs of unstable modes $(m,n) = (1,\pm 2)$, $(2,\pm 1)$, $(2,\pm 2)$, $(1,\pm 3)$. The corresponding amplitudes are denoted by ξ_1, ξ_2, ξ_3, ξ_4, respectively. The structure of the order parameter equations may easily be established on the basis of the invariance condition (3.8). They take the form:

$$\partial_t\, \xi_1 = \lambda_1\xi_1 - C_{10}\xi_2^*\xi_4 - \xi_1 \sum_k C_{1k}|\xi_k|^2 - C_{15}\xi_2\xi_3^*\xi_4 - C_{16}\xi_1^*\xi_3^2$$

$$\partial_t\, \xi_2 = \lambda_2\xi_2 - C_{20}\xi_1^*\xi_4 - \xi_2 \sum_k C_{2k}|\xi_k|^2 - C_{25}\xi_1\xi_3\xi_4^*$$

$$\partial_t\, \xi_3 = \lambda_3\xi_3 \qquad\qquad - \xi_3 \sum_k C_{3k}|\xi_k|^2 - C_{35}\xi_1^*\xi_2\xi_4 - C_{36}\xi_1^2\xi_3^*$$ (4.13)

$$\partial_t\, \xi_4 = \lambda_4\xi_4 - C_{40}\xi_1\xi_2 - \xi_4 \sum_k C_{4k}|\xi_k|^2 - C_{45}\xi_1\xi_2^*\xi_3$$

λ_k, C_{jk} are complex, $j,k=1,..4$.

The coefficients have been determined from the basic equation. As a necessary condition, $\text{Re}(C_{kk})>0$, otherwise the trajectories can leave any finite region in phase space. This condition turns out to be fulfilled for the quasi-geostrophic two-layer model [13].

174

Fig. 4: Baroclinic instability: A typical chaotic evolution of flow pattern;
after [13]

Depending on the control parameters, there exists a variety of attractors of
the order parameter equations. The competition of modes may lead to a state
described only by one pair of modes. In this case we again obtain a rotating
wave solution. A simple eddy pattern, as shown in Fig. 3 , propagates with
constant shape and constant phase velocity around the sphere. By changing
control parameters, states can be found which consist of a superposition of two
modes. This may be a superposition of two rotating waves. Quasi-periodic
behaviour in addition to frequency-locked states have been found. A competition
of four modes shows a transition from quasi-periodic behaviour to a chaotic
dynamics, see Fig. 4.

5. Waves in Systems with O(2)-Symmetry

A system with O(2)-symmetry possesses, in addition to the SO(2)-symmetry, an
invariance with respect to the inversion $\phi \rightarrow -\phi$ of the cyclic variable ϕ.
An example of a physical system exhibiting O(2)-symmetry, is the ring laser.
Another quite large class of systems, which can, to some extent, be discussed in
the context of oscillatory instabilities in systems with O(2)-symmetry, are
systems with a spatially homogeneous state undergoing instabilities induced by
modes varying periodically in space on a scale much smaller than the size of the
system. In this case, one approximates the system by an infinitely extended
system and imposes the condition of spatial periodicity. As typical examples, we
cite the oscillatory onset of convection in binary fluid mixtures in long and
narrow rectangular containers and the Taylor-Couette experiment with counter-
rotating cylinders. The description, in terms of an infinitely extended system
with spatial periodicity, introduces an artificial O(2)-symmetry, which is
broken in actual systems. Quite interesting phenomena result from this broken
symmetry which will be discussed below.

175

Due to the O(2)-symmetry, the normal modes again have the form

$$\Phi_{jm}(\underline{r}) = \Phi_{jm}(r,z) \exp(im\phi) \quad . \tag{5.1}$$

As a result of the inversion symmetry $\phi \to -\phi$, the modes with $-m$ have the same eigenvalue λ_{jm} as the modes with m. Therefore, there are at least four unstable modes with the same real part of the corresponding eigenvalue:

$$\Phi_{jm}(\underline{r}), \ \Phi_{j-m}(\underline{r}), \ \Phi_{jm}(\underline{r})^{*}, \ \Phi_{j-m}(\underline{r})^{*} \quad . \tag{5.2}$$

If we assume that only one group of these four modes become unstable, then the state vector has the following shape close to instability:

$$\underline{q}(\underline{r},t) = \underline{q}^0(r,z) + \xi_\ell(t) \ \Phi_{um_u}(r,z) \exp\{im_u\phi\}$$
$$+ \ \xi_r(t) \ \Phi_{u-m_u}(r,z) \exp\{-im_u\phi\} + c.c. \tag{5.3}$$
$$+ \ \underline{w}(\xi_\ell,\xi_r,r,\phi,z) \quad .$$

Here, $\underline{w}(\xi_\ell,\xi_r,r,\phi,z)$ denotes the contribution of the stable modes, which is at least quadratic in ξ_1, ξ_r. The order parameter equations for the amplitudes ξ_ℓ and ξ_r have to be invariant with respect to the transformations (translational symmetry)

$$\xi_\ell \to \xi_\ell \exp\{im_u\phi_0\}, \qquad \xi_r \to \xi_r \exp\{-im_u\phi_0\}, \tag{5.4}$$

as well as (inversion symmetry)

$$\xi_\ell \to \xi_r \ ; \ \xi_r \to \xi_\ell \quad . \tag{5.5}$$

They therefore have the general form

$$d_t \ \xi_\ell = (\varepsilon + i\omega_0) \ \xi_\ell - \xi_\ell \ (\ A \ |\xi_\ell|^2 + B \ |\xi_r|^2 \ )$$
$$d_t \ \xi_r = (\varepsilon + i\omega_0) \ \xi_r - \xi_r \ (\ A \ |\xi_r|^2 + B \ |\xi_\ell|^2 \ ) \quad , \tag{5.6}$$

if an approximation of the nonlinearity by polynomials up to the third order in ξ_ℓ, ξ_r is sufficient. Additionally, terms like $\xi_\ell^2\xi_r$, which are not excluded by symmetry arguments, have been dropped by using the rotating wave approximation, since they vary on a much faster time scale ($\approx \exp(3i\omega_0 t)$) compared to ξ_ℓ, ξ_r ($\approx \exp(i\omega_0 t)$). We mention that for the case of the Taylor-Couette experiment, which in this kind of description posseses O(2)×SO(2)-symmetry, these terms vanish identically due to the additional SO(2)-rotational symmetry in the azimuthal direction.

The behaviour of a system with O(2)-symmetry undergoing oscillatory instability can be easily classified by using the the order parameter equations (5.6). They admit two different types of time periodic solutions:

$$a) \ \xi_\ell = (\varepsilon/\text{Re}(A))^{1/2} \exp\{i\omega t\}, \ \xi_r = 0$$
$$\xi_r = (\varepsilon/\text{Re}(A))^{1/2} \exp\{i\omega t\}, \ \xi_\ell = 0, \tag{5.7}$$
$$\omega = \omega_0 - \text{Im}(A)/\text{Re}(A) \ \varepsilon$$

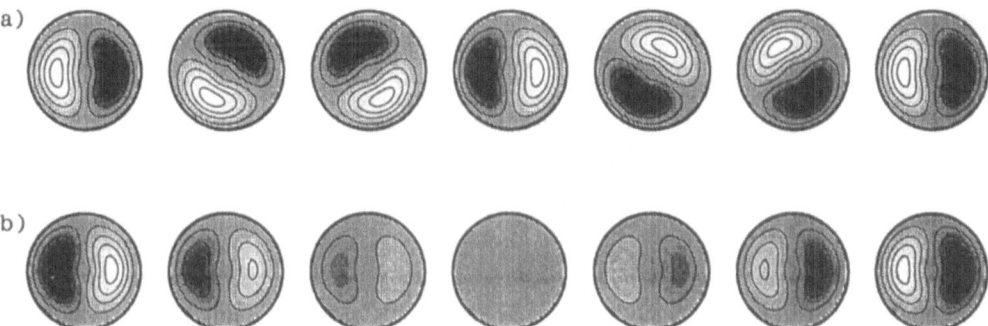

Fig. 5: The basic spatio-temporal patterns arising due to the oscillatory
instability in the presence of O(2)-symmetry (Wave number m=1):
a) Counterclockwise rotating wave
b) Standing wave

$$\text{b) } \xi_\ell = \xi_r = (\varepsilon/Re(A+B))^{1/2} \exp\{i\omega t\}$$

$$\omega = \omega_0 - Im(A+B)/Re(A+B) \; \varepsilon \qquad .$$

(5.8)

(We have assumed Re(A)>0, Re(A+B)>0, A≠B). A stability analysis shows that
for Re(A)/Re(B)<1 the solutions a) are stable. For Re(A)/Re(B)>1 the solution b)
turns out to be stable, the case Re(A)=Re(B) is degenerate and higher order
terms have to be included into the nonlinearities of the order parameter
equations. An inspection of the spatial structure of the two types of solutions
shows that type a) corresponds to a wave running either in the ϕ- direction or
in the opposite direction. Type b) describes a standing wave solution, see
Fig. 5. That these types of behaviour exhaust all possibilities may be seen by
deriving the following set of differential equations for the absolute values
r_r, r_ℓ of ξ_ℓ, ξ_r:

$$d_t \; r_\ell = -\frac{d}{dr_\ell} \; V(r_r, r_\ell) \quad , \quad d_t \; r_r = -\frac{d}{dr_r} \; V(r_r, r_\ell)$$

(5.9)

$$V(r_r, r_\ell) = -\varepsilon/2 \; (r_r^2 + r_\ell^2) + A/4 \; r_r^4 + B/2 \; r_r^2 r_\ell^2 + A/4 \; r_\ell^4$$

The absolute values of the order parameters behave just like an overdamped
particle moving in the potential $V(r_r, r_\ell)$. In the Taylor-Couette system with
counter-rotating cylinders, standing waves as well as travelling waves are
observed experimentally. They are known as 'ribbons' and 'spirals', respectively
[14]. Convection in binary mixtures takes the form of travelling waves.

6. Waves in Systems with O(3)-Symmetry

Instabilities of a spherically symmetric state can also lead to wave-like
behaviour. Due to the O(3)-symmetry of these systems, the behaviour close to
instability can again be investigated, in principle, by establishing the order
parameter equations and considering their solutions. Here, the classification of
the various types of behaviour is far from complete. We only present certain
examples.

The normal modes of the linear eigenvalue problem arising in the examination
of the stability of the spherically symmetric state (for the case of a scalar
quantity $T(r,\phi,\theta)$) have the form

$$T(r,\phi,\theta) = T_{j\ell}(r) \; Y_\ell^m(\phi,\theta) \qquad .$$

(6.1)

They can be classified with respect to their transformation properties under rotations. The corresponding eigenvalues are at least $(2\ell+1)$-fold degenerate due to symmetry. Various situations and instabilities may arise. One may consider instabilities of one group of modes belonging to the ℓ-th irreducible representation for the oscillatory or nonoscillatory case or instabilities of groups of modes belonging to different irreducible representations of the $O(3)$-group. The corresponding order parameter equations can again be derived from symmetry considerations.

First, we may consider an instability with a single group of modes belonging to the ℓ-th irreducible representation of the $O(3)$-group (modes eq. (6.1) characterized by ℓ). Here, the instability can be nonoscillatory (real eigenvalue) or oscillatory (complex eigenvalue). The nonoscillatory instability leads to stationary patterns. However, the oscillatory instability again gives rise to standing and travelling waves, and more complicated types of behaviour could also be expected, especially for large values of ℓ. Let us briefly discuss the oscillatory instability of the group of modes with $\ell=1$. The order parameter equations take the form (supposing an expansion up to terms of third order is sufficient)

$$d_t \, \xi_m = \lambda \, \xi_m - \alpha \, I_2 \, (-1)^m \frac{\partial}{\partial \xi_{-m}} \, I_1 - \beta \, I_1 \, (-1)^m \frac{\partial}{\partial \xi_{-m}} \, I_1$$

$$I_1 = 2 \, \xi_1 \xi_{-1} - \xi_0 \xi_0 \tag{6.2}$$

$$I_2 = \xi_1 \xi_1^* + \xi_0 \xi_0^* + \xi_{-1} \xi_{-1}^* \quad .$$

They depend, on the linear eigenvalue λ, and on two coefficients α and β. There is always a stable solution; either a standing wave pattern or a rotating wave pattern. The analysis of the oscillatory instability due to modes with higher values of ℓ becomes rather involved and has only been partly accomplished.

Wave-like behaviour has also been found close to nonoscillatory instabilities induced by two groups of modes belonging to the irreducible representation with $\ell=1$ and $\ell=2$. Due to mode interaction various types of time dependent behaviour may arise. This has been demonstrated for a specific model describing the onset of convection in a spherical fluid layer as a result of the Bénard instability [15]. The model consists of a fluid contained between two spherical boundaries in a spherically symmetric gravitational field. A radially symmetric temperature gradient induces, just as in the planar Bénard experiment, an unstable density distribution which leads to the onset of convection. The model describes convection in the earth's mantle. Other geophysical applications are fairly obvious.

The order parameter equations for the instability of a spherical symmetric state in the case of mode interaction with modes with $\ell=1$ and $\ell=2$ can again be established by group theoretic methods [15]. They take the following qualitative form:

$$d_t \, \xi_m = \lambda_1 \, \xi_m + (-1)^m \frac{\partial}{\partial \xi_{-m}} \, \{ \, \alpha \, Q^{(3)}(\xi:\xi:\eta) + a \, E(\xi:\xi)^2 +$$

$$b \, E(\xi:\xi) \, E(\eta:\eta) + c \, Q^{(4)}(\xi:\xi:\eta:\eta) \, \}$$

$$d_t \, \eta_m = \lambda_2 \, \xi_m + (-1)^m \frac{\partial}{\partial \eta_{-m}} \, \{ \, \beta \, Q^{(3)}(\eta:\eta:\eta) + \delta \, Q^{(3)}(\eta:\eta:\eta) +$$

$$d \, E(\eta:\eta)^2 + e \, E(\xi:\xi) \, E(\eta:\eta) + f \, Q^{(4)}(\xi:\xi:\eta:\eta) \, \} \quad .$$

(6.3)

The order parameters ξ_m and η_m are the amplitudes of the groups of modes with

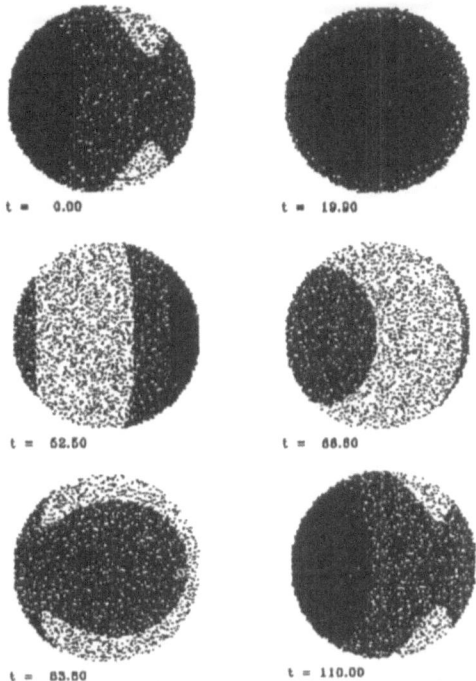

Fig. 6: Convection in spherical shells: Temperature field of rotating wave
motion generated by the competition of modes with $\ell=1$ and $\ell=2$;
after [15]

$\ell=1$ and $\ell=2$, respectively. $E(\xi:\xi)$, $E(\eta:\eta)$, $Q^{(3)}(\xi:\xi:\eta)$, $Q^{(3)}(\eta:\eta:\eta)$, and $Q^{(4)}(\xi:\xi:\eta:\eta)$ are polynomial invariants of the $O(3)$-group (for a definition see [15]). The general form of the order parameter equations holds for different systems with $O(3)$-symmetry, which exhibit a simultaneous instability of modes with $\ell=1$ and $\ell=2$. Consideration of a concrete system specifies the values of the coefficients $\alpha,..$,a, b,.., which have to be calculated from the basic equations. These equations show an interesting temporal behaviour in the form of waves. Figure 6 shows a time periodic rotating wave motion. Quasiperiodic motions are obtained as well as chaotic behaviour. A sequence of patterns in the case of chaotic behaviour is exhibited in Fig. 7. The patterns are spatially coherent, since they consist of modes with $\ell=1$ and $\ell=2$. Nevertheless, the temporal evolution is rather complex.

7. Waves with Cellular Structures in Largely Extended Systems

Instabilities of a spatially homogeneous state in largely extended systems quite often lead to structures with a spatial periodicity much smaller than the size of the system. A well-known pattern, for example, is the roll structure of the convection pattern of the Bénard instability. The first Bénard instability exhibits a non-oscillatory character and results in a stationary pattern. Currently, great effort is being devoted to the investigation of the behaviour of the cellular structure in the case of an oscillatory instability. Typical systems are convection in binary mixtures [16] and waves in the Taylor-Couette experiment with counterrotating cylinders. As suggested previously, a rough theoretical approach consists in considering the behaviour of an infinitely extended system and imposing a spatial periodicity. This turns the problem into

179

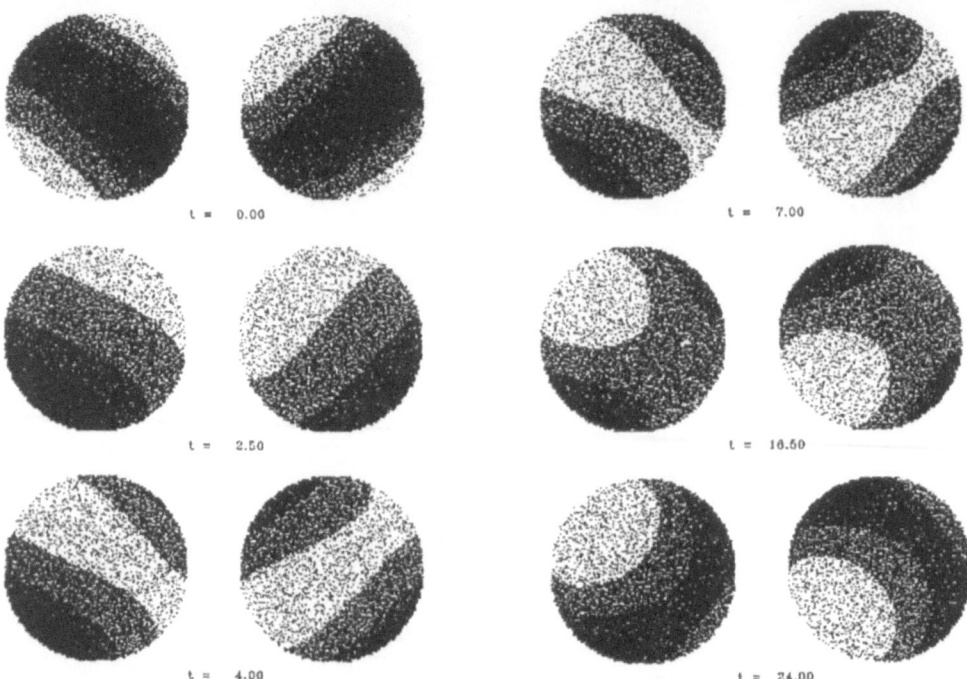

Fig. 7: Convection in spherical shells: Temperature field of chaotic wave
motion generated by the competition of modes with $\ell=1$ and $\ell=2$
in systems with $O(3)$-symmetry; after [15]

the investigation of an oscillatory instability in a system with $O(2)$-symmetry,
and we know that the behaviour is described by the order parameter equations
(5.6) and that, typically, standing waves or running waves will emerge.

In the following we extend this rough theoretical description into a
treatment which allows us to include the effects of the boundaries as well as
variations in the spatial structure. Nevertheless, the instability can be
described by an order parameter equation which is, to some extent, independent
of the concrete nature of the system. We consider systems with two different
length scales. Examples are the Taylor-Couette system with large cylinders and a
binary fluid mixture in a container with large horizontal dimensions compared to
the thickness of the fluid layer. We present the results obtained for the latter
case.

Consider, for instance, the geometry of the convection instability. Horizontal coordinates are denoted by $\underline{x}$, the vertical coordinate is denoted by z.
There are continuously many modes characterized by the continuous two-dimensional
wave number $\underline{k}$. Modes with different spatial variations with respect to the
coordinate z are distinguished by the discrete index j.

$$\Phi_{j\underline{k}}(\underline{r}) = \Phi_{j\underline{k}}(z) \exp\{i \ \underline{k}\underline{x}\} \quad . \tag{7.1}$$

We consider an oscillatory instability, therefore the eigenvalues $\lambda_{j\underline{k}}$ are
complex. Due to symmetry, they are degenerate and depend only on the absolute
value $|\underline{k}|$:

$$\lambda_{j\underline{k}} = \lambda_j(\underline{k}^2) \quad . \tag{7.2}$$

180

Since $\underline{k}^2$ can vary continuously in the infinitely extended system, there is, close to instability, a band of unstable modes, i. e. there are uncountably many unstable modes. One may self-consistently assume that at the onset of instability, only modes close to the most unstable one (wave number $|\underline{k}|_c$) will be excited. The modes corresponding to a stable band of eigenvalues are damped and are enslaved by the modes of the unstable band. Therefore the modes corresponding to the stable bands can be eliminated adiabatically. The solution is constructed as a superposition of the unstable modes and a smaller contribution $\underline{w}$ of the stable modes, which is a functional of the amplitudes $\xi_{\underline{k}}(t)$ of the unstable modes:

$$\underline{q}(\underline{r},t) = \int d\underline{k}\ \xi_{\underline{k}}(t)\ \Phi_{u\underline{k}}(z)\ \exp\{i\underline{k}\underline{x}\} + c.c. + \underline{w}((\xi_{\underline{k}}))\quad . \tag{7.3}$$

We obtain the set of order parameter equations

$$d_t\ \xi_{\underline{k}}(t) = \lambda_u(\underline{k}^2)\ \xi_{\underline{k}}(t) +$$

$$\int d\underline{k}_1\ d\underline{k}_2\ d\underline{k}_3\ a^3(\underline{k};\underline{k}_1,\underline{k}_2,\underline{k}_3)\ \xi_{\underline{k}_1}(t)\ \xi_{\underline{k}_2}(t)\ \xi_{\underline{k}_3}(t)^* \quad . \tag{7.4}$$

We have dropped terms proportional to $\xi_{\underline{k}_1}(t)\ \xi_{\underline{k}_2}(t)$ and $\xi_{\underline{k}_1}(t)\ \xi_{\underline{k}_2}(t)\ \xi_{\underline{k}_3}(t)$ by performing a rotating wave approximation as discussed in Sect. 6. To obtain the order parameter equation in a more elegant form, we introduce the order parameter field

$$\xi(\underline{x},t) = \int d\underline{k}\ \xi(\underline{k},t)\ \exp\{i\underline{k}\underline{x}\} \tag{7.5}$$

to obtain the equation

$$d_t\ \xi(\underline{x},t) = \lambda_u(-\Delta)\ \xi(\underline{x},t) +$$

$$\int d\underline{x}_1\ d\underline{x}_2\ d\underline{x}_3\ a^3(\underline{x};\underline{x}_1,\underline{x}_2,\underline{x}_3)\ \xi(\underline{x}_1,t)\ \xi(\underline{x}_2,t)\ \xi(\underline{x}_3,t)^* \quad . \tag{7.6}$$

Here, Δ denotes the two-dimensional Laplacian. Since we expect that only modes with absolute values of the wave number $\underline{k}$ close to the critical one are excited, we can approximate the eigenvalue $\lambda_u(\underline{k}^2)$ by

$$\lambda_u(\underline{k}^2) = \gamma\varepsilon +\lambda_0 - \lambda_1\ \underline{k}^2 + \lambda_2\ \underline{k}^4 \tag{7.7}$$

in such a way that the expression is correct close to $\underline{k}_c^2$. Furthermore, it is, possible under certain conditions, to approximate the nonlocal nonlinear interaction term by a local one. This has been shown for the case of convection in binary mixtures for a certain set of boundary conditions [17]. In this case, one ends up with an order parameter equation of the following form:

$$d_t\ \xi(\underline{x},t) = (\gamma\varepsilon +\lambda_0 + \lambda_1\ \Delta + \lambda_2\ \Delta^2)\ \xi(\underline{x},t) + \alpha\ |\xi(\underline{x},t)|^2\ \xi(\underline{x},t)^* +$$

$$\beta\ |\nabla\xi(\underline{x},t)|^2\ \xi(\underline{x},t)^* + \gamma\ \xi(\underline{x},t)^*\ (\nabla\xi(\underline{x},t))^2 \quad . \tag{7.8}$$

The state vector $\underline{q}(\underline{r},t)$ has the structure

$$\underline{q}(\underline{r},t) = \Phi_{u(\underline{k}=-i\nabla)}(z)\ \xi(\underline{x},t) + c.c. + \underline{w}((\xi(\underline{x},t))) \quad . \tag{7.9}$$

From this expression one can derive, at least in an approximate form, boundary conditions for the order parameter field $\xi(\underline{x},t)$.

8. Waves in Large Aspect Ratio Systems: Convection in Binary Mixtures

The order parameter equation (7.8) has been used to examine the onset of oscillatory convection in binary fluid mixtures, e.g. in ethanol-water mixtures. Let us briefly describe the main characteristics of the convection instability in such a mixture [16]. The fluid is contained between two horizontal plates and is heated from below and cooled from above. The density of a binary mixture depends both on temperature as well as on the relative concentrations of the two components of the fluid. If a fluid particle is removed from its initial position upwards in the vertical direction, its density is smaller compared to the surrounding fluid and it starts to move further upwards. The temperature gradient across the layer therefore induces an instability by the same mechanism as in the ususal Bénard experiment. But, since the density of a binary fluid depends on the concentration ratio of its two components, a temperature gradient between the fluid particle and its surroundings will lead to a change of the concentration ratio due to the Soret effect and thus to a change of density. This may result in an increase in the density so that the fluid particle falls back down to its initial position from where it is pushed upwards again. The fluid particle therefore undergoes an oscillatory motion, in contrast to the usual Bénard experiment. The behaviour of the system is described in mathematical form by the Navier-Stokes equation for the velocity field $\underline{v}(\underline{r},t)$, the equation of heat diffusion and a diffusion equation for the relative concentration $C(\underline{r},t)$:

$$P^{-1}[\partial_t \underline{v}(\underline{r},t) + (\underline{v}(\underline{r},t)\nabla)\underline{v}(\underline{r},t)] =$$

$$\Delta\underline{v}(\underline{r},t) + \underline{e}_3 R[\theta(\underline{r},t) - \Psi C(\underline{r},t)] - \text{grad } p(\underline{r},t) \quad,$$

$$\text{div } \underline{v}(\underline{r},t) = 0 \quad, \tag{8.1}$$

$$\partial_t \theta(\underline{r},t) + (\underline{v}(\underline{r},t)\nabla)\theta(\underline{r},t) = v_3(\underline{r},t) + \Delta\,\theta(\underline{r},t) \quad,$$

$$\partial_t C(\underline{r},t) + (\underline{v}(\underline{r},t)\nabla)\,C(\underline{r},t) = -v_3(\underline{r},t) + L\,\Delta[\theta(\underline{r},t)+C(\underline{r},t)] \quad,$$

where P denotes the Prandtl number, L the Lewis number and Ψ is the separation ratio, which are all material constants of the fluid mixture. The Rayleigh number R is a dimensionless measure for the temperature difference across the layer and serves as a control parameter. The vector $\underline{e}_3$ denotes the unit vector in the vertical direction. The system has been investigated for the case of free boundary conditions for the velocity field and fixed relative concentration $C(\underline{r},t)$ on the upper and lower plates. To this end, the coefficients arising in the order parameter equation (7.8) have been determined from the basic equations (8.1). They are given in [17].

Let us first discuss results obtained by a numerical integration of the order parameter equation for a narrow rectangular geometry. The convective state, which develops as a result of the instability of the purely heat conductive state, is a state consisting of two wave-trains consisting of left travelling waves on the left and right travelling waves on the right side of the container, Fig. 8. It is quite interesting to note that the basic flow does not consist of one travelling wave, as suggested by the usual treatment introducing spatial periodicity. Apparently, the topology of the states occurring in finite systems is entirely different. For a slightly higher value of the temperature difference across the layer, the pattern changes to a time periodic pattern where one of the two wave parts dominates. A further increase of the temperature gradient leads to a quasiperiodic flow, which consists of a right travelling wave motion on the right vertical boundary and a left travelling wave on the opposite side. The intensities of these wave trains vary periodically in time, Fig.9. We

182

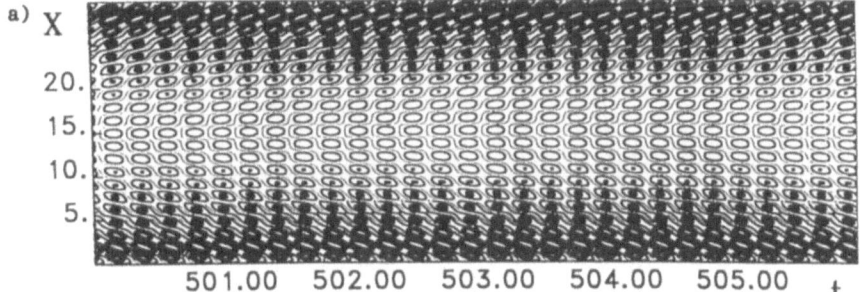

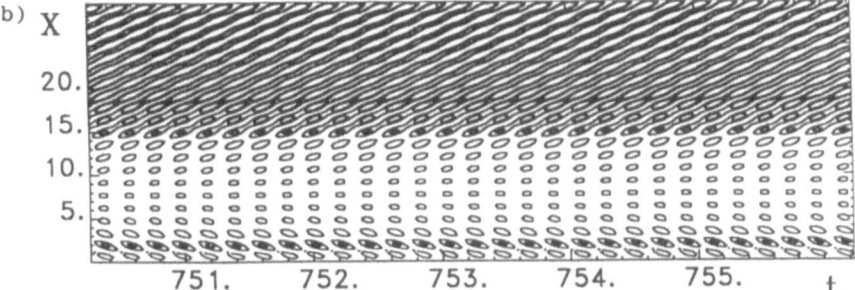

Fig. 8: Convection in binary mixtures: Lines of constant vertical velocity as
functions of time and the horizontal x-coordinate for a narrow
rectangular container (aspect ratio: 30).
   a) A flow consists of two symmetric waves travelling towards the
      sidewalls
   b) A flow consists of one asymmetric wave travelling towards a
      sidewall; after [17]

mention that this quasiperiodic flow has already been observed in experiments on
convection in binary fluids [16] and has been termed the 'blinking state'. For
higher values of the Rayleigh number, the flow calculated by the order parameter
equation becomes chaotic. The patterns almost consist, for a while, of a right
or left running wave on the right or left side of the container and then change
to one consisting of a wave propagating in the opposite direction. These changes
apparently form a random sequence, as can be seen in Fig. 10.

Two-dimensional wave patterns have also been obtained by a numerical
investigation of the order parameter equation [18]. In the case of rectangular
containers, the rolls of the travelling waves are aligned almost perpendicular to
the larger side of the container. Waves travelling to the right are located on
the right side and waves travelling to the left on the opposite side. Close to
the onset of convection, the spatial structure of the rolls is only slightly
disturbed, but a transverse instability of the rolls is clearly indicated, in
accordance with experimental findings. For higher values of the Rayleigh number,
the waves show the tendency to be confined to localized regions, Fig. 11. These
'confined states' are also obtained experimentally. Typical patterns are
exhibited in Fig. 12. Two-dimensional wave patterns in large aspect ratio
systems with circular geometry are shown in Figs. 13. Close to the onset of
convection, the patterns consist of wave-trains moving in radial directions. The
rolls are more or less deformed transversely and modulated in time. For higher
values of the Rayleigh number, the convective rolls are confined to a small ring
at the circular boundary. They are aligned perpendicular to the boundary and
move in the azimuthal direction. The dynamics of the wave patterns is chaotic.

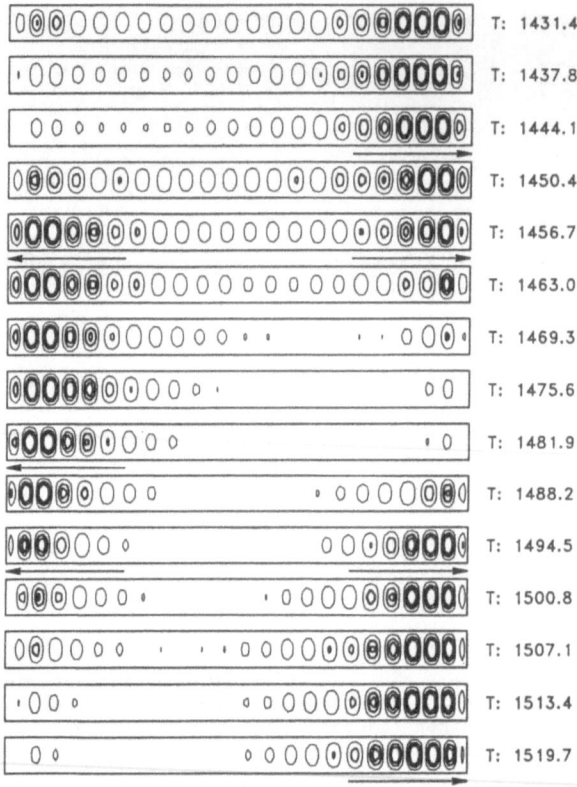

	T: 1431.4
	T: 1437.8
	T: 1444.1
	T: 1450.4
	T: 1456.7
	T: 1463.0
	T: 1469.3
	T: 1475.6
	T: 1481.9
	T: 1488.2
	T: 1494.5
	T: 1500.8
	T: 1507.1
	T: 1513.4
	T: 1519.7

Fig. 9: Convection in binary mixtures: 'Blinking state'. The amplitudes
of the two wave trains travelling towards the sidewalls are alternating.
The fluid motion is quasiperiodic

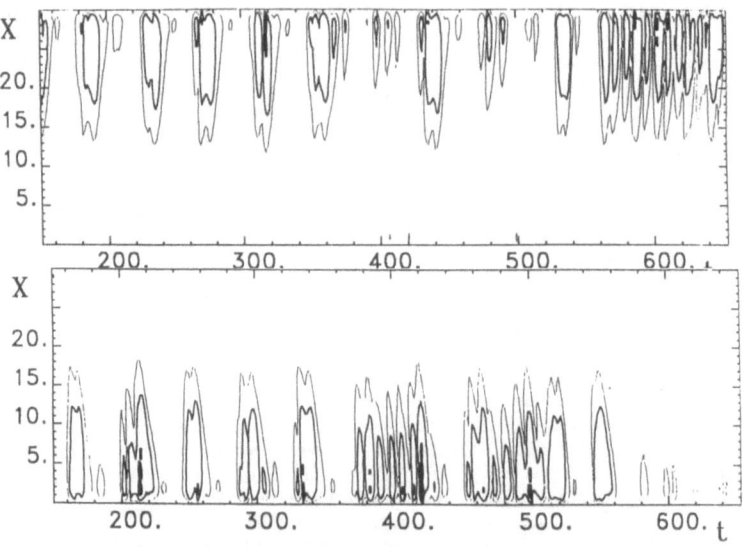

Fig. 10: Convection in binary mixtures: Slowly varying envelope functions of
left-travelling (top) and right-travelling waves (bottom) if the fluid
motion is chaotic; after [17]

a)

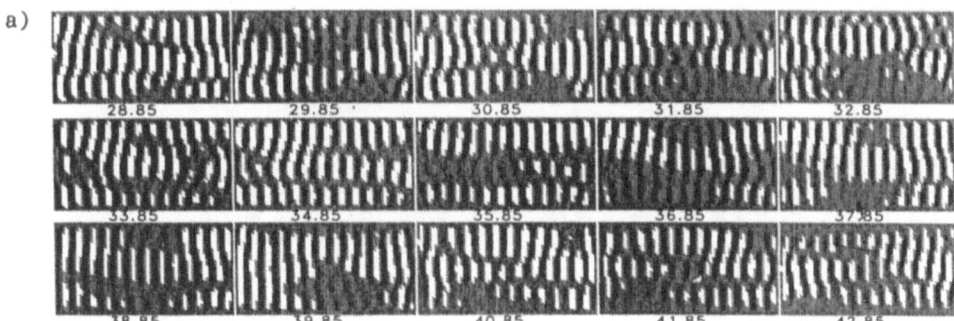

b)

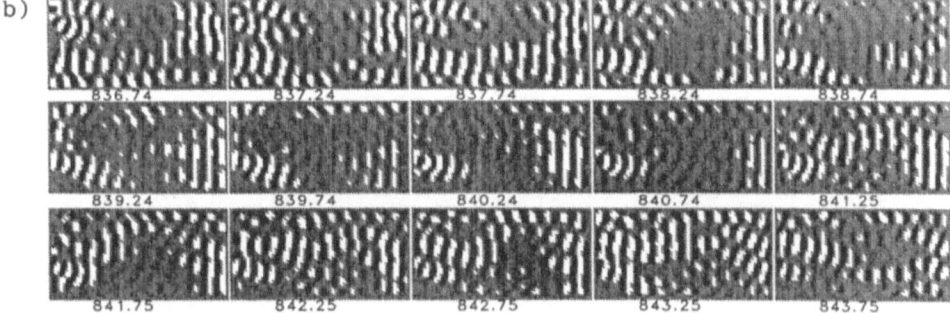

Fig. 11: Convection in binary mixtures: Travelling wave patterns in a rectangular container.
a) Situation close to threshold: 'Zipper state'. The waves travel towards the sidewalls exhibiting a transverse instability.
b) Situation well above threshold: 'Confined states'.
The waves are localized in confined regions;
after [18]

In summary, we see that waves with cellular structures have a tendency to exhibit irregular behaviour in space as well as in time. Due to the universality of the order parameter equation, similar spatio-temporal behaviour as that discussed for binary fluid mixtures is expected to occur in quite different systems which exhibit an oscillatory instability associated with a cellular spatial structure.

9. Waves in a Biological System: α-EEG Waves

The occurrence of wave-like behaviour is not restricted to physical and chemical systems. Waves are also encountered in biological systems. A well-known example is provided by waves of the electric potential produced by the activity of neurons. These waves are measurable in the human skull in forms of encephalograms (EEG). The electric potential derived between two different locations exhibits more or less sinusoidal and regular temporal oscillations with a dominant frequency of about 8-13 Hz. The oscillations in this frequency range have been termed 'α-rhythm' and are measurable while the person is in a quiescent state and has closed eyes. The α-rhythm is not specific for the human brain, it is also measurable in animals. The EEG of the human brain is produced by the action of about 10^{11} neurons and the fact that one observes more or less regular temporal oscillations is a hint that self-organization takes place, correlating the behaviour of neurons at different locations in the brain. As is well known, patterns of a synergetic system can be analyzed on a

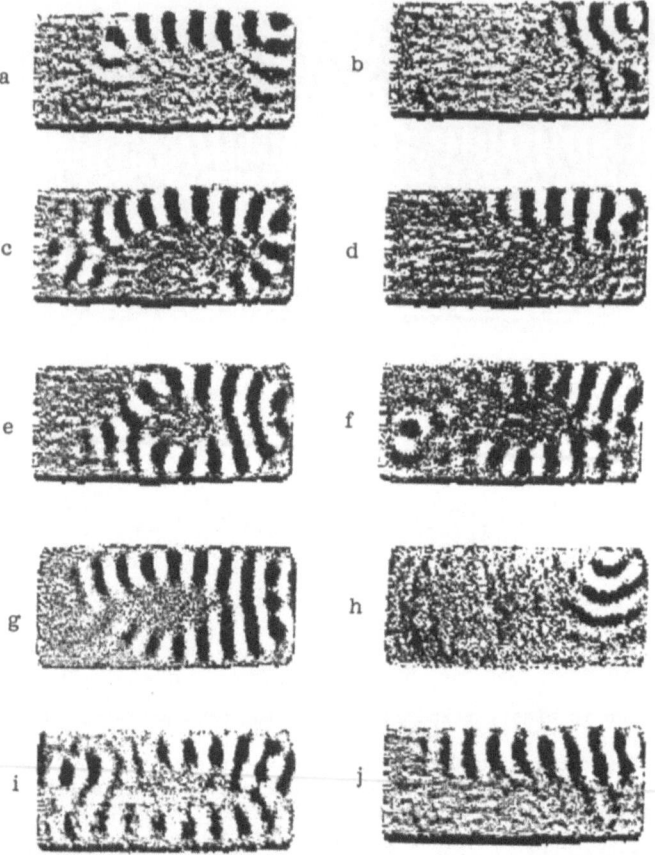

Fig. 12: Experimentally obtained travelling wave patterns in rectangular geometry, after [16]

macroscopic level without detailed knowledge of the subsystems composing the whole system. This suggests that an analysis of the waves should be possible without considerations of the behaviour of single neurons. The concepts of synergetics seem to be the accurate tools to analyze and understand the basic features of the dynamics of the EEG-activity, which are the macroscopic patterns produced by the cooperative behaviour of the neuronal subsystems.

In the following, we discuss an analysis of a data set of α-EEG waves, experimentally derived by Lehmann and his coworkers, in accordance with the general lines of synergetics. The data we analyze were recorded in 16 channels from a normal male volunteer of 35 years of age, who was seated in a sound-shielded chamber. The array scheme is shown in Fig. 14. The electrode at the vertex was used as a common reference. The time series of the signal of each channel shows, besides the characteristic 10 Hz oscillations, a modulation of the signal on a larger time scale. This modulation of the oscillations is apparently stochastic and it has been speculated that the underlying dynamics is a chaotic process [19]. However, more insight is gained by considering the spatio-temporal patterns. In order to analyze the patterns they were mapped onto a circular planar region. Although the time signal for each channel is rather complex, the dynamics of the spatial patterns of the electric potential are apparently coherent. The patterns at each time step consist mainly of two regions of different polarity. Less frequently, two minima or maxima are

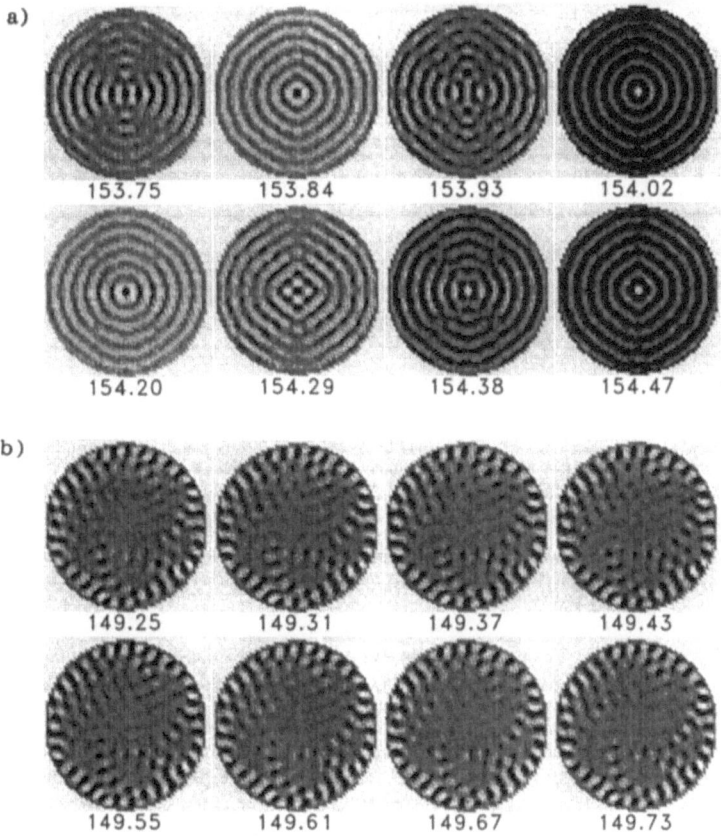

a)

153.75 153.84 153.93 154.02

154.20 154.29 154.38 154.47

b)

149.25 149.31 149.37 149.43

149.55 149.61 149.67 149.73

Fig. 13: Convection in binary mixtures: Travelling wave patterns in a
circular box.
a) Situation close to threshold: Waves travel more or less in radial
direction exhibiting a transverse instability.
b) Situation well above threshold: The waves are confined to the circular
boundary and travel in counterclockwise direction;
after [18]

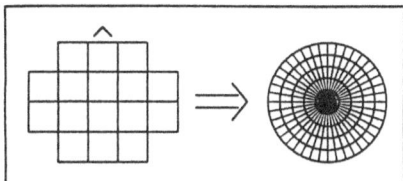

Fig. 14: Array scheme of the electrodes and the mapping of the derived
patterns onto a circular region; after [20]

observed. The patterns undergo, roughly speaking, two different kinds of motion.
Usually, the pattern rotates clockwise or anticlockwise in the azimuthal
direction. Thereby, the pattern is only slightly deformed, Fig. 15. Apart from
this deformation, it resembles a rotating wave motion. The frequency of the
rotation is approximately 10 Hz. The time intervals between the clockwise and

Fig. 15: Temporal evolution of α-EEG patterns: Rotating wave pattern

Fig. 16: Temporal evolution of α-EEG patterns: Standing wave pattern

the counterclockwise rotation of the pattern are random. Less frequently, one observes a qualitatively different behaviour, where the rotation of the pattern ceases and the polarity reverses periodically, Fig. 16: Regions of positive and negative polarity exchange. Roughly speaking, the motion resembles a standing wave pattern. The jumps between the clockwise and counterclockwise rotation and the oscillation are related to the modulation of the 10 Hz oscillation of the time signal of one channel.

The spatial patterns observed and their dynamics remind us strongly of the wave motions discussed previously in physical systems. It is therefore tempting to analyze the spatio-temporal behaviour along the lines indicated by the synergetic theory of self-organization. Let us briefly summarize the main results of this analysis (see [20] for a more detailed description). The EEG-patterns were projected onto a circular planar region. The patterns given by the electric potential $U(\underline{r},t)$ were developed into a complete set of modes of the form

$$\Phi_{jm}(r,\phi) = \Phi_{jm}(r) \exp im\phi$$

(9.1)

$$U(r,\phi,t) = \sum_{jm} \xi_{jm}(t)\, \Phi_{jm}(\underline{r})$$

where r denotes the radial direction and ϕ measures the angle in the azimuthal direction. By choosing a suitable basis $\Phi_{jm}(r,\phi)$, it was possible (see [20]) to explicitly demonstrate that each pattern of the total time series may be well represented by the modes with

$$\Phi_{1,0}(r,\phi),\ \Phi_{1,\pm1}(r,\phi),\ \Phi_{2,\pm2}(r,\phi) \quad .$$

(9.2)

All other modes do not contribute significantly. In fact, the reconstruction of the patterns by the above modes is accurate up to more than 97 per cent. It therefore seems tempting to interpret the amplitudes of these modes as order parameters. Additionally, the close analogy to the behaviour of waves in physical systems, especially travelling and standing waves in systems with O(2)-symmetry, (see Sect. 5.), suggests that the dynamics of the α-EEG patterns may be due to nonlinear mode interaction and is therefore governed by a set of order parameter equations. However, the dynamics of these order parameters has still to be analyzed in more detail. Furthermore, there are clearly small scale structures varying on a much smaller time-scale. For a successful interpretation of the spatio-temporal patterns on the basis of order parameters, these degrees of freedom need to act as small fluctuations and do not significantly influence the dynamics on large scales.

a)

b)

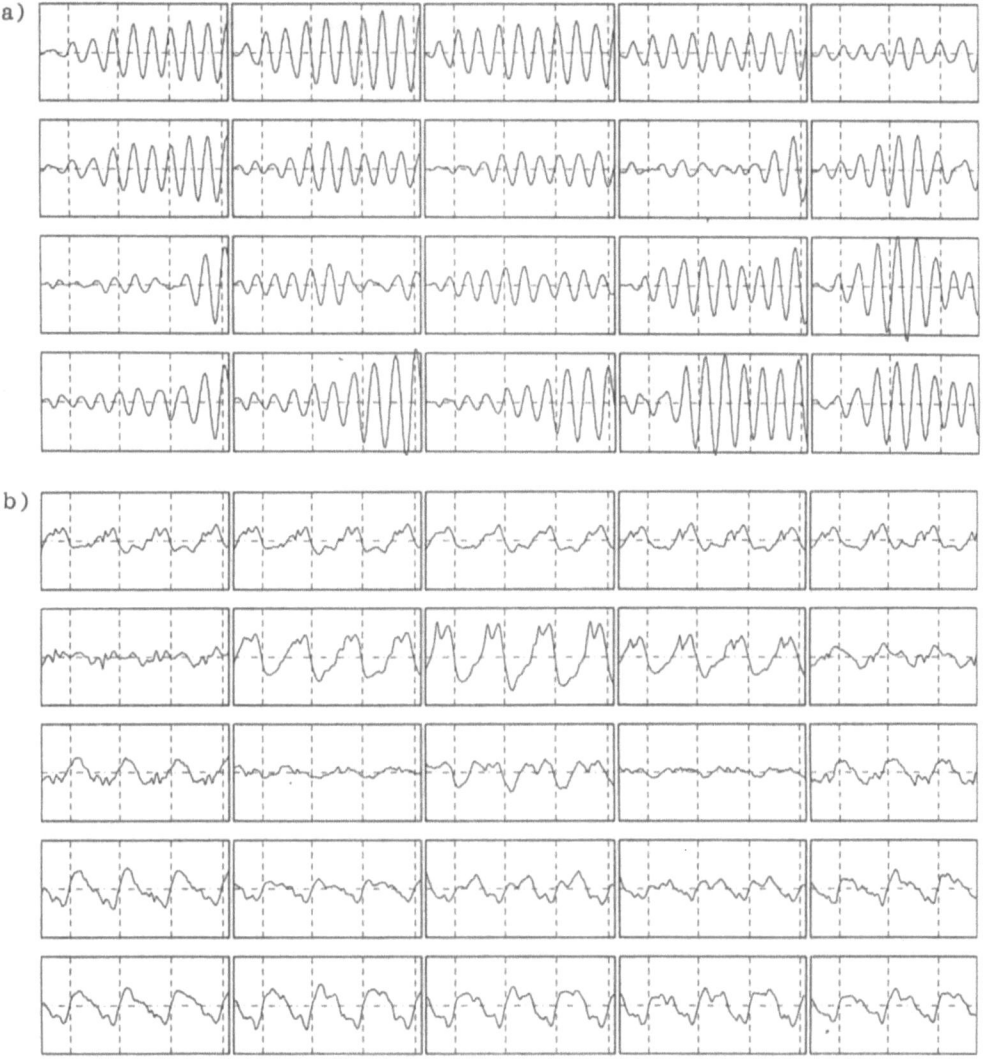

Fig. 17: Time signals of the EEG potential difference between two points
of the skull.
a) α waves
b) Signal during an epileptic fit

A strong suggestion that these modes may in fact be interpreted as order parameters, lies in the examination of EEG patterns derived from a person during an epileptic fit (The data set has again been derived by Lehmann and coworkers). As is well-known, the time signal of one channel is more or less time-periodic, Fig. 17. The corresponding evolution of the spatial patterns is exhibited in Fig. 18. The main change lies in the fact that the rotating wave motion has ceased and the pattern again consists of two regions of different polarity and the polarity reverses periodically in time. The analysis of the patterns, along the same lines as for the 'normal EEG' patterns, yields the following, quite remarkable result: The patterns are again dominated by the modes with azimuthal wave numbers m=0, m=±1, m=±2, but in the epileptic state the dynamics of the

189

Fig. 18: A period of the time-periodic EEG patterns during an epileptic fit.
The temporal development starts on the bottom line left and ends on
the upper right corner

modes m=0 and m=±2 turns out to be rigidly coupled. It seems that this rigid
coupling between the modes is the characteristic feature of the epileptic EEG
pattern. The fact that the dynamics of the EEG-patterns in the epileptic state
can be described by the same set of modes as the patterns in a normal state is a
hint that the dynamics is produced by the nonlinear interaction between the
amplitudes of modes. The transition to an epileptic state could then be
interpreted as a transition from a chaotic state to a time periodic behaviour by
a rigid coupling of degrees of freedom in the space of the order parameters.

10. Associative Memory and Pattern Recognition

It was emphasized already some years ago [21] that there is a close analogy
between pattern formation and pattern recognition. This analogy has been
exploited in the development of a dynamical model for pattern recognition,
devised by H. Haken [22]. This model has been implemented on a computer and
works successfully as an associative memory. This has explicitly been
demonstrated by the reconstruction of faces only parts of which are offered to
the system and which then are restored by the dynamical process [23].

Let us first set up the model for pattern recognition. To this end we
consider pattern formation in a synergetic system, for instance oscillatory
instability in systems with O(2)-symmetry as discussed above. We assume the case
of an instability for stable travelling waves. If we start with an initial
condition which is a superposition of the two waves running in different
directions the system will eventually evolve into a state consisting of just one
travelling wave. This selective property of the system is quite interesting from
the point of view of the problem of pattern recognition. A device can be used to
perform pattern recognition if it is able to identify an offered pattern with
one of a set of prototype patterns. Systems undergoing an oscillatory
instability in the case of O(2)-symmetry exhibit this property for the case of
two prototype patterns in the following sense: If we start with a state vector

$q(\underline{r},0)$, the system evolves into one of two possible running wave states. This happens in such a way that the wave, which initially had the largest overlap with the offered state $q(\underline{r},0)$, is selected. We may identify the initial state vector with an offered pattern which is eventually recognized by the system in the final state of a left or right running wave, depending on the initial overlap of the offered pattern and prototype pattern. We demonstrate in the following how this model can be constructed in close analogy to the instability of travelling waves above discussed. This emphasizes the close relationship between the processes of pattern formation and pattern recognition.

We consider a system which has the task of recognizing two different patterns which we denote by two state vectors $\Phi_1(\underline{x})$, $\Phi_2(\underline{x})$ representing for instance, two different images in a two-dimensional plane. We define a state vector

$$q(\underline{x},t) = \xi_1(t) \, \Phi_1(\underline{x}) + \xi_2(t) \, \Phi_2(\underline{x}) + w(\underline{x},t) \ . \tag{10.1}$$

The initial pattern consists of a superposition of the protoptype patterns and an additional contribution $w(\underline{x},t)$. For the following, it is necessary to introduce adjoint prototype patterns by the relations

$$<\Phi_i^+(\underline{x})|\Phi_j(\underline{x})> = \int d^2\underline{x} \ \Phi_i^+(\underline{x}) \ \Phi_j(\underline{x}) = \delta_{ij} \tag{10.2}$$

where we assume that the adjoint vectors can be expressed as a linear combination of $\Phi_i(\underline{x})$:

$$\Phi_i(\underline{x})^+ = \sum_j g_{ij} \ \Phi_j(\underline{x}) \qquad . \tag{10.3}$$

These two conditions define the adjoint vectors uniquely, if the prototype patterns are linearly independent. In close analogy with the treatment of pattern formation, we assume that the vector $w(\underline{x},t)$ is orthogonal to the prototpye vectors,

$$<\Phi_i^+(\underline{x})|w(\underline{x},t)> = 0 \qquad , \tag{10.4}$$

and we identify $w(\underline{x},t)$ as a superposition of stable modes and $\xi_i(t)$ as order parameters. To construct the model for pattern recognition, we introduce a dynamical system for the state vector $q(\underline{x},t)$ with the following properties:

1) The vector $w(\underline{x},t)$ has to vanish in the course of time:

$$\lim_{t\to\infty} ||w(\underline{x},t)|| \to 0 \qquad . \tag{10.5}$$

2) The amplitudes $\xi_i(t)$ (i=1,2) have to fulfill the dynamical system

$$d_t \, \xi_1(t) = \lambda\xi_1(t) - \xi_1(t) \, (\, A \, \xi_1(t)^2 + B \, \xi_2(t)^2 \,)$$

$$d_t \, \xi_2(t) = \lambda\xi_2(t) - \xi_2(t) \, (\, A \, \xi_2(t)^2 + B \, \xi_1(t)^2 \,) \tag{10.6}$$

with $A/B < 1$, $A > 0$, $B > 0$

which is analogous to the order parameter equations for the amplitudes of the running waves for the oscillatory instability in the presence of $0(2)$-symmetry. The state vector $q(\underline{x},t)$ then obeys the evolution equation

$$d_t \, q(\underline{x},t) = [\lambda - (\, A \, \xi_1(t)^2 + B \, \xi_2(t)^2 \,)]\xi_1(t) \quad \Phi_1(\underline{x})$$

$$+ [\lambda - (\, A \, \xi_2(t)^2 + B \, \xi_1(t)^2 \,)]\xi_2(t) \quad \Phi_2(\underline{x}) \; + d_t \, w(\underline{x},t) \qquad . \tag{10.7}$$

We can represent the amplitudes $\xi_i(t)$ by the scalar products

$$\xi_i(t) = <\Phi_i(\underline{x})^+ | q(\underline{x},t)> \tag{10.8}$$

to obtain the following form of the evolution equation:

$$d_t\, q(\underline{x},t) = \Sigma_{i=1,2}\ [\ \Phi_i(\underline{x})<\Phi_i(\underline{x})^+|q(\underline{x},t)>$$

$$\times\ [\ \lambda - B\ \Sigma_{j\neq i} <\Phi_j(\underline{x})^+|q(\underline{x},t)>^2\ ]\ ]$$

$$- A\ \Sigma_{i=1,2}\ \Phi_i(\underline{x})\ <\Phi_i(\underline{x})^+|q(\underline{x},t)> <\Phi_i(\underline{x})^+|q(\underline{x},t)>^2 +d_t w(\underline{x},t) \tag{10.9}$$

Now we are free to choose an appropriate dynamics for the vector $w(\underline{x},t)$ which has to satisfy condition (10.5). We choose the dynamics

$$d_t w(\underline{x},t) = -\ A\ (\ \xi_1(t)^2 + \xi_2(t)^2\ )\ w(\underline{x},t) \tag{10.10}$$

which obviously has the desired limiting behaviour for large values of t and we obtain the model equation:

$$d_t\, q(\underline{x},t) = \Sigma_{i=1,2}\ [\ \Phi_i(\underline{x})<\Phi_i(\underline{x})^+|q(\underline{x},t)>$$

$$\times\ [\ \lambda - (B-A)\ \Sigma_{j\neq i} <\Phi_j(\underline{x})^+|q(\underline{x},t)>^2\ ]\ ]$$

$$- A\ q(\underline{x},t)\ \Sigma_{i=1,2} <\Phi_i(\underline{x})^+|q(\underline{x},t)>^2\ . \tag{10.11}$$

This equation may readily be extended to the case of N prototype patterns. From the construction of the dynamical system it is evident that the system reaches the state

$$\lim_{t\to\infty} q(x,t) = \xi_k\ \Phi_k(\underline{x}) \tag{10.12}$$

which consists of the prototype pattern $\Phi_k^+(\underline{x})$, which initially had the largest overlap

$$\int d^2\underline{x}\ \Phi_k^+(\underline{x})\ q(\underline{x},0) \tag{10.13}$$

with the offered pattern $q(\underline{x},0)$. The dynamical system is therefore able to retrieve an initially deformed pattern. Furthermore, it can restore an incomplete pattern and act as an associative memory.

The algorithm for pattern recognition has been successfully implemented on a computer. The prototype patterns consist of faces and labels which are capital letters in the upper righthand corner of the patterns, Fig. 19. Now, either the label can be offered to the system as an initial condition $q(\underline{x},0)$ and the dynamical system (10.11) recalls the face (see Fig. 20), or a part of the face can be offered and the dynamical system performs a completion of the whole face including the corresponding label. The dynamical system acts as a content-addressable associative memory. Furthermore, it has been shown that the above model of pattern recognition can also be successfully applied to the recognition of shifted, rotated and scaled patterns. To this end, linear preprocessing of the patterns is performed, which turn them into patterns which are invariant with respect to translations, rotations and scalings. For further discussion we refer the reader to [23].

Fig. 19: Pattern recognition: The set of stored prototype patterns; after [23]

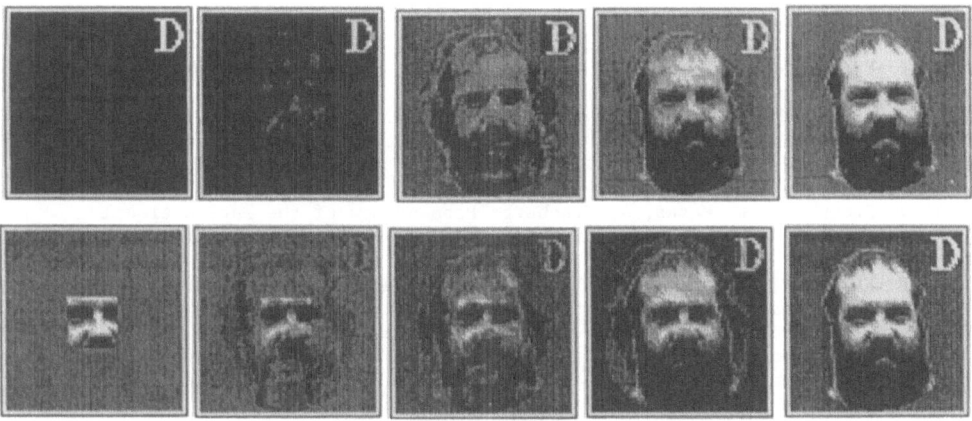

Fig. 20: Pattern recognition: Time evolution during the associative recall
for two different initial conditions; after [23]

In summary, we see that the ideas and methods which are applied successfully
in the investigation of pattern formation in systems far from equilibrium, are
also of fundamental importance in pattern recognition. Pattern recognition can
be viewed as the dual process of pattern formation. The pattern recognition
algorithm presented above has been constructed in close analogy to the
mechanisms underlying pattern formation in nonequilibrium systems. It demonstrates that specific pattern recognition tasks may be cast in a mathematical
form. Furthermore, due to the close analogy between pattern formation and
pattern recognition, one may try to find ways of fulfilling these tasks using
synergetic systems. For example, the large class of systems with O(2)-symmetry
which undergo an oscillatory instability, may serve as a realization of the
dynamical process of pattern recognition for the case of two prototype patterns.
Finally, we mention that by suitably changing the dynamical system (10.11) for
the amplitudes of the protopype patterns, one can construct dynamical systems
which model other phenomena occuring in pattern recognition such as hysteresis
and oscillations in perception [24].

References

[1] H. Haken: "Synergetics. An Introduction" 3rd. edition, Springer,
    Berlin, Heidelberg 1983
[2] H. Haken: "Advanced Synergetics", Springer, Berlin, Heidelberg 1983
[3] E. Basar, H. Flohr, H. Haken, A.J. Mandell (Eds.): "Synergetics of the
    Brain", Springer, Berlin, Heidelberg 1984
[4] A. Wunderlin, H. Haken: in "Synergetics - From Microscopic to Macroscopic
    Order", Ed. E. Frehland, Springer, Berlin, Heidelberg 1982
[5] W. Weidlich, G. Haag: "Concepts and Models of Quantitative Sociology"
    Springer, Berlin, Heidelberg 1982
[6] A. Wunderlin, H. Haken: Z. Phys. $\underline{B44}$, 135, (1981)
[7] H. Haken, A. Wunderlin: Z. Phys. $\underline{B47}$, 179, (1982)
[8] M. Bestehorn: Dissertation Universität Stuttgart (1988)
[9] D. H. Sattinger: "Group Theoretic Methods in Bifurcation Theory",
    Lecture Notes in Mathematics Vol. 762, Springer, Berlin, Heidelberg 1979
[10] H. L. Swinney and J.P. Gollub: "Hydrodynamic Instabilities and the
    Transition to Turbulence", Topics in Applied Physics Vol. 45
    Springer, Berlin, Heidelberg 1982
[11] K. Marx and H. Haken: Europhys. Lett. $\underline{5}$, 315, (1988);
    K. Marx: Dissertation Universität Stuttgart (1987)
[12] N.A. Phillips: J. Meteor. $\underline{8}$, 391, (1951); Tellus $\underline{6}$, 273, (1954)
[13] H. Haken, W. Weimer: Z. Geomorph. N. F., Suppl.-Bd. $\underline{67}$, 103, (1988)
[14] W. F. Langford, R. Tagg, E. J. Kostelich, H. Swinney, and M. Golubitsky:
    Phys. Fluids $\underline{31}$, 776, (1988)
[15] R. Friedrich, H. Haken: Phys. Rev. $\underline{A34}$, 2100, (1986);
    R. Friedrich, H. Haken: in "The Physics of Structure Formation",
    edited by W. Güttinger and G. Dangelmayr,
    Springer Berlin, Heidelberg 1987
[16] V. Steinberg, E. Moses, J. Fineberg: Proceedings of the International
    Conference on "The Physics of Chaos and Systems Far From Equilibrium",
    Nucl. Phys. $\underline{B2}$, 109, (1987);
    P. Kolodner, A. Passner, H. L. Williams, C. M. Surko: Proceedings of the
    International Conference on "The Physics of Chaos and Systems
    Far From Equilibrium", Nucl. Phys. $\underline{B2}$, 97, (1987)
[17] M. Bestehorn, R. Friedrich, and H. Haken: Z. Phys. $\underline{B72}$, 265, (1988)
[18] M. Bestehorn, R. Friedrich, and H. Haken: Z. Phys. $\underline{B}$, to appear (1989)
[19] A. Babloyantz: Strange Attractors in the Dynamics of Brain Activity,
    in "Complex Systems - Operational Approaches", ed. by H. Haken,
    Springer, Berlin, Heidelberg 1985
[20] A. Fuchs, R. Friedrich, H. Haken, and D. Lehmann: Spatio-Temporal Analysis
    of Multichannel α-EEG Map Series, in "Computational Systems -
    Natural and Artificial", ed. by H. Haken, Springer, Berlin, Heidelberg 1987
[21] H. Haken: Pattern Formation and Pattern Recognition - An Attempt at a
    Synthesis, in "Pattern Formation by Dynamical Systems and
    Pattern Recognition", ed. by H. Haken, Springer, Berlin, Heidelberg 1979
[22] H. Haken: Computers for Pattern Recognition and Associative Memory,
    in "Computational Systems - Natural and Artificial", ed. by
    H. Haken, Springer, Berlin, Heidelberg 1987
[23] A. Fuchs and H. Haken: Biol. Cybern. $\underline{60}$, 17, (1988);
    Biol. Cybern. $\underline{60}$, 107, (1988)
[24] H. Haken: Synergetics in Pattern Formation and Associative Action,
    in "Neural and Synergetic Computers", ed. by H. Haken,
    Springer Berlin, Heidelberg 1988

Index of Contributors